Behavior Analysis and Modeling of Traffic Participants

Synthesis Lectures on Advances in Automotive Technology

Editor
Amir Khajepour, *University of Waterloo*

The automotive industry has entered a transformational period that will see an unprecedented evolution in the technological capabilities of vehicles. Significant advances in new manufacturing techniques, low-cost sensors, high processing power, and ubiquitous real-time access to information mean that vehicles are rapidly changing and growing in complexity. These new technologies—including the inevitable evolution toward autonomous vehicles—will ultimately deliver substantial benefits to drivers, passengers, and the environment. Synthesis Lectures on Advances in Automotive Technology Series is intended to introduce such new transformational technologies in the automotive industry to its readers.

Cyber-Physical Vehicle Systems: Methodology and Applications
Chen Lv, Yang Xing, Junzhi Zhang, and Dongpu Cao
2020

Reinforcement Learning-Enabled Intelligent Energy Management for Hybrid Electric Vehicles
Teng Liu
2019

Deep Learning for Autonomous Vehicle Control: Algorithms, State-of-the-Art, and Future Prospects
Sampo Kuuti, Saber Fallah, Richard Bowden, and Phil Barber
2019

Narrow Tilting Vehicles: Mechanism, Dynamics, and Control
Chen Tang and Amir Khajepour
2019

Dynamic Stability and Control of Tripped and Untripped Vehicle Rollover
Zhilin Jin, Bin Li, and Jungxuan Li
2019

Real-Time Road Profile Identification and Monitoring: Theory and Application
Yechen Qin, Hong Wang, Yanjun Huang, and Xiaolin Tang
2018

Noise and Torsional Vibration Analysis of Hybrid Vehicles
Xiaolin Tang, Yanjun Huang, Hong Wang, and Yechen Qin
2018

Smart Charging and Anti-Idling Systems
Yanjun Huang, Soheil Mohagheghi Fard, Milad Khazraee, Hong Wang, and Amir Khajepour
2018

Design and Avanced Robust Chassis Dynamics Control for X-by-Wire Unmanned Ground Vehicle
Jun Ni, Jibin Hu, and Changle Xiang
2018

Electrification of Heavy-Duty Construction Vehicles
Hong Wang, Yanjun Huang, Amir Khajepour, and Chuan Hu
2017

Vehicle Suspension System Technology and Design
Avesta Goodarzi and Amir Khajepour
2017

Behavior Analysis and Modeling of Traffic Participants

Xiaolin Song and Haotian Cao

ISBN: 978-3-031-00381-3   paperback
ISBN: 978-3-031-01509-0   PDF
ISBN: 978-3-031-00013-3   hardcover

DOI 10.1007/978-3-031-01509-0

A Publication in the Springer series
SYNTHESIS LECTURES ON ADVANCES IN AUTOMOTIVE TECHNOLOGY

Lecture #15
Series Editor: Amir Khajepour, *University of Waterloo*
Series ISSN
Print 2576-8107   Electronic 2576-8131

Behavior Analysis and Modeling of Traffic Participants

Xiaolin Song and Haotian Cao
Hunan University, China

SYNTHESIS LECTURES ON ADVANCES IN AUTOMOTIVE TECHNOLOGY #15

ABSTRACT

A road traffic participant is a person who directly participates in road traffic, such as vehicle drivers, passengers, pedestrians, or cyclists, however, traffic accidents cause numerous property losses, bodily injuries, and even deaths to them. To bring down the rate of traffic fatalities, the development of the intelligent vehicle is a much-valued technology nowadays. It is of great significance to the decision making and planning this vehicle if the pedestrians' intentions and future trajectories, as well as those of surrounding vehicles, could be predicted, all in an effort to increase driving safety. Based on the image sequence collected by onboard monocular cameras, we use the Long Short-Term Memory- (LSTM) based network with an enhanced attention mechanism to realize the intention and trajectory prediction of pedestrians and surrounding vehicles.

However, although the fully automatic driving era still seems far away, human drivers are still a crucial part of the road–driver–vehicle system under current circumstances, even dealing with low levels of automatic driving vehicles. Considering that more than 90% of fatal traffic accidents were caused by human errors, thus it is meaningful to recognize the secondary task while driving, as well as the driving style recognition, to develop a more personalized advanced driver assistance system (ADAS) or intelligent vehicle. We use the graph convolutional networks for spatial feature reasoning and the LSTM networks with the attention mechanism for temporal motion feature learning within the image sequence to realize the driving secondary-task recognition.

Moreover, aggressive drivers are more likely to be involved in traffic accidents, and the driving risk level of drivers could be affected by many potential factors, such as demographics and personality traits. Thus, we will focus on the driving style classification for the longitudinal car-following scenario. Also, based on the Structural Equation Model (SEM) and Strategic Highway Research Program 2 (SHRP 2) naturalistic driving database, the relationships among drivers' demographic characteristics, sensation seeking, risk perception, and risky driving behaviors are fully discussed. Results and conclusions from this short book are expected to offer potential guidance and benefits for promoting the development of intelligent vehicle technology and driving safety.

KEYWORDS

intelligent vehicle, trajectory prediction, pedestrian intention prediction, secondary driving task recognition, long short-term memory network with attention mechanism, car-following driving style classification, naturalistic driving data analysis, risky driving behavior, structural equation model

Contents

Acknowledgments

This work was supported by the National Natural Science Foundation of China (grant numbers 51975194 and 51905161) and the Hunan Provincial Natural Science Foundation (grant numbers 2021JJ30121 and 2021JJ40067).

The authors would like to thank Mr. Yanbin Zeng, Mr. Huijie Shi, Mr. Chaopeng Pan, Mr. Yangang Yin, and Ms. Xuemei Jia for their kind help while writing this short book. We are also thankful to Morgan & Claypool Publishers for providing the opportunity and support for this book.

Xiaolin Song and Haotian Cao
October 2021

CHAPTER 1

Introduction

A Road traffic participant is a person who directly participates in road traffic, such as vehicle drivers, passengers, pedestrians, or cyclists, however, traffic accidents cause numerous property losses, bodily injuries, and even deaths to them. According to a report by the World Health Organization in 2018 [1], the traffic accident is the 8th leading cause of death for people of all ages. The number of road traffic deaths remains unacceptably high. More than 1.35 million people, including pedestrians, cyclists, and motorcyclists, die in traffic accidents every year. Therefore, it is meaningful to take action to improve traffic safety for all traffic participants. Over the last decade, the intelligent vehicle has received unprecedented attention from domains of research institutes, industries, and governments, which is expected to be a hopeful solution for reducing traffic accidents, while also achieving the ultimate driving comfort, safety, and efficiency. The intelligent vehicle integrates multiple key modules such as environment perception, localization, fusion, decision-making, planning, and motion control, many advanced functionalities such as pedestrian protection systems (PPS), automatic emergency braking systems (AEBS), advanced driver assistance systems (ADASs), and even high-automatic driving systems can be seamlessly integrated. To prevent potential traffic accidents, the decision-making and planning subsystems of the intelligent vehicle are needed to accurately understand the driving environment, while the dynamic changes of the moving intentions and trajectories of the pedestrians and surrounding vehicles bring major challenges.

Moreover, it should be noted that human and behavioral factors play key roles in road traffic accidents [2]. Existing reports indicate that the percentage of crashes involving certain types of driving errors made by the driver was as high as 94% [3]. More than 90% of fatal traffic accidents were caused by human errors [1], such as dring, fatigue, distraction, and mal-operation, even in the era of automatic driving. According to the definition of driving automation from the Society of Automotive Engineers International (SAE International), human drivers in level-2 or level-3 automated-driving vehicles are allowed to participate in secondary tasks (e.g., texting, phone talking) while driving [4], but the driver still has the responsibility to take over the vehicle in emergencies. Thus, it's necessary to monitor the driver's activities and evaluate the driver's take-over ability to assist automated driving systems to take proper actions.

Besides, different types of human driving behavior occurrences may be related to *psychological* factors. Previous studies have linked drivers' personality traits, psychological factors, and driving behaviors to crash participation [5–7]. Drivers' deliberate driving violations and unintentional mistakes are, respectively, associated with the more unemotional and more impulsive

Table 1.1: Methods of vehicle trajectory prediction

Method	Complexity	Factors Considered	Challenges
Motion-based methods	Low	The vehicle dynamical or vehicle kinematical characteristics	Sensitive to the initial state and sensor noises
Maneuver-based methods	Middle	Maneuver features, vehicle's dynamic, or kinematics characteristics	Un-observability, subjectivity, the complexity of maneuvers
Interaction-aware methods	High	Vehicle information, road structure, and traffic rules	Interaction is difficult to detect and model

aspects of psychologies [8]. Drivers with a greater number of anger traits are more likely to be provoked and tend to engage in more aggressive behaviors [9]. Therefore, the driver's demographics, personality traits, and driving behaviors play crucial roles in traffic safety. Exploring relationships between driver's demographics, personality traits, and risky driving behaviors could be helpful in identifying specific factors related to driving safety. Meanwhile, these factors can be further utilized to identify high-risk drivers, such that it is expected to reduce property losses, injuries, and even death caused by traffic accidents.

In a word, this book will talk about the state-of-art regarding the behavior modeling and prediction of traffic participants, more specifically, the intention and trajectory prediction of the surrounding vehicles and pedestrians, the secondary driving task recognition, car-following driving styles, and the relationships among the risky driving behaviors and human factors, which is expected to offer potential guidance and benefits for promoting the development of intelligent vehicle technology and driving safety. For a better understanding of the contents in this book, a literature review regarding these topics will be simply given in the following contents.

1.1   TRAJECTORY PREDICTION OF THE VEHICLE

As shown in Table 1.1, vehicle trajectory prediction methods generally can be divided into three categories according to complexity level, which includes the motion-based methods, maneuver-based methods, and interaction-aware trajectory prediction models.

1.1.1   MOTION-BASED TRAJECTORY PREDICTION METHODS

There are two types of vehicle motion models that are usually adopted, namely, the dynamic model and the kinematic model. The dynamic model is based on the Lagrange equation to describe the motion of the vehicle, which considers the influence of different types of forces such as the lateral and longitudinal force of the tire, cornering force acting on the vehicle [10]. The

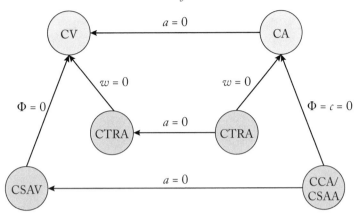

Figure 1.1: Relationship diagram of different kinematic models.

complete dynamic model needs to consider the engine and transmission parameters of the vehicle, which is complicated and inconvenient, and it is more suitable for vehicle control. Therefore, the simple bicycle model is more widely used for vehicle trajectory prediction [11]. Barnstorm et al. [12] used a linear bicycle model to predict the trajectory of the host vehicle and obstacles in the future, and proposed a decision-making method to avoid collisions for the host vehicle drive. Lin et al. [13] predicted the future trajectory of the host vehicle based on the bicycle model and steady-state Kalman filter to realize some safety functions (such as pedestrian collision warning). In addition, there are many related studies [14–17] that use dynamic models to predict vehicle trajectories.

While the kinematics model does not consider the influence of force, it describes the movement of the vehicle based on the mathematical relationship between the vehicle's motion state (such as position, speed, angular velocity, acceleration, yaw angle). Schubert et al. [18] conducted a review study on kinematics models. As shown in Figure 1.1, the simplest kinematic models are the constant velocity (CV) model and the constant acceleration (CA) model [19–23]. The CV and CA models are mainstream models which assume that the speed or acceleration of the vehicle holds constant, respectively. The constant turning rate and speed (CTRV) model and the constant turning rate and acceleration (CTRA) model are quadratic models, which take into account the movement in the Z-axis direction and introduce the yaw angle and angular velocity to describe the curved motion of the vehicle [20, 22, 24–27]. The constant steering angle and velocity (CSAV) model and the constant curvature and acceleration (CAA) model convert the angular velocity in CTRV and CTRA to the wheel angle, thereby realizing the coupling between vehicle velocity and angular velocity.

The above-mentioned models can be used to predict the future trajectory of the target vehicle but still without uncertainty considered. The direct prediction method is the simplest one without considering the uncertainty which assumes the sensor information to be accurate

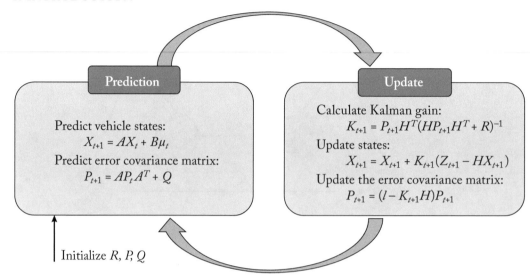

Figure 1.2: Kalman filtering process.

and could detect the required vehicle states. The direct prediction method does not perform additional processing on the uncertainty and generates a certain trajectory, which can be applied to dynamic models [12] or kinematics models [21, 23, 27]. This method generally has high computational efficiency and is suitable for applications with strong real-time requirements. However, the prediction reliability could be limited because it does not consider sensor errors and model errors.

Kalman filter and its variant models are widely used to deal with the uncertainty of predicted trajectories. It assumes that the noise of the sensor and the model obey Gaussian distributions. The Kalman filter [28] could estimate the state of the vehicle recursively based on the measurements with noise. As shown in Figure 1.2, in the prediction phase, the state X_{t+1} and the error covariance matrix P_{t+1} at the next time step are estimated. While in the update phase, the Kalman gain matrix K_{t+1} is calculated first, and then the vehicle state X_{t+1} is updated with measurements, and finally the error covariance matrix P_{t+1} is updated. When performing trajectory prediction, the predicted trajectory coordinates and corresponding variances of the target vehicle can be output through multiple iterations during the prediction stage, to process the uncertainty [33, 34].

However, Kalman Filter approximates the noise by a single-peak Gaussian distribution that is not completely accurate. Some studies adopted switching the Kalman filter [37, 38] to improve it. Overall, the motion-based trajectory prediction method has the advantages of simplicity, efficiency, and high real-time performance, but only considers the low-level attributes of vehicle motion (e.g., vehicle motion states), thus, the long-term prediction accuracy might be unsatisfactory.

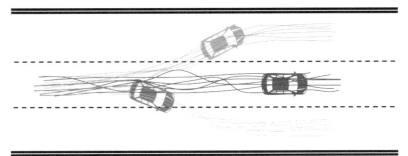

Figure 1.3: Cluster of vehicle trajectories on a structured road.

1.1.2   MANEUVER-BASED TRAJECTORY PREDICTION MODELS

The maneuver-based trajectory prediction methods recognize the driving maneuver of the vehicle first and then generate the corresponding predicted trajectory, which performs better in the long-term prediction window when compared with the motion-based methods. We will introduce the methods based on prototype trajectory and maneuver recognition, respectively.

Methods based on the prototype trajectory: As shown in Figure 1.3, these methods are based on the prototype trajectory assuming that the trajectories of vehicles on the structured road can be divided into a set of finite trajectory clusters, while each cluster corresponds to a motion mode (or saying maneuver) of the vehicle. The known trajectory is used to match the most likely motion mode, and then the corresponding prototype trajectory is used to predict its future trajectory, and the Gaussian Process is a typical and popular one [31–34], which assumes that the trajectories in the training set are the samples of the Gaussian Process, then fits each cluster with a Gaussian distribution. During the phase of predictions, given a known historical trajectory, it is necessary to calculate the distance between this trajectory and each motion mode, where the probability [31–33], average Euclidean distance [35], improved Hausdorff distance [36], maximum common subsequence [37], and other indicators could be used. After matching the most likely motion mode through the above indicators, the prototype trajectory corresponding to the motion mode can be directly used as the predicted trajectory [35]. It can also calculate the probability that the target vehicle belongs to different motion modes at the current moment, thereby outputting multiple predicted trajectories with weights [38].

Methods Based on Maneuver Recognition: Methods based on maneuver recognition identify the current maneuver of the target vehicle and generate the predicted trajectory accordingly. Compared with prototype trajectory-based methods, methods based on maneuver recognition can use higher-level features such as speed and acceleration to more accurately recognize the maneuver. Many classification methods [39–42] can be used to recognize the driving maneuver of the target vehicle. Once the maneuver is identified, there still needs to generate corresponding predicted trajectories according to the maneuver. For example, Tamke et al. [43] deduced the corresponding control values according to the maneuver and imported them into a kinematic model to predict trajectories. Houenou et al. [44] used splines of different maneuvers to output the fine predicted trajectories, in which the experimental results show that the prediction error during 0–1 s and 3–4 s are 0.09 m and 0.45 m, respectively. Aoude et al. [45] sampled and generated the RRT tree in the input space of the motion model, and then predicted the trajectories of obstacles to realize a collision warning to improve driving safety.

1.1.3   INTERACTION-AWARE TRAJECTORY PREDICTION MODELS

Still, the maneuver-based trajectory prediction method has certain limitations because it does not consider the interaction between the target vehicle and its surrounding environment. While the interaction-aware trajectory prediction models take the vehicle as an interactive entity, who interacts with its surrounding environment such as the vehicles, roads, traffic rules. These methods are generally divided into two categories, namely, methods of explicit modeling interaction and methods of implicit modeling interaction.

Methods of Explicit Modeling Interaction: Explicit modeling interaction methods model the interaction between vehicles explicitly in a logically understandable way. Studies in [46] and [47] use collision risk among predicted trajectories to model the interaction between vehicles. Brand et al. [48] used the Coupled Hidden Markov Models (CHMMs) to construct the coupled dependency relationship among multiple vehicles. Agamennoni et al. [49, 50] used a traffic rule-based method to construct the interaction among vehicles. In addition, game theory [51–53] is widely introduced in this field, which regards vehicles as game entities and seeks equilibrium in a multi-party game traffic environment.

Methods of Implicit Modeling Interaction: Methods of implicit modeling interaction are commonly used in neural networks to describe the interaction among vehicles. But the internal interaction details of the learned vector are difficult to be understood from a logical perspective. For instance, Alahi et al. [54] built a Social-Long Short-term Memory (LSTM) model based on the LSTM neural network and described the interaction by constructing social interaction vectors. Experimental results show that this method achieved the best results on multiple datasets such as the ZARA1, the ZARA2, and the UCY. Deo et al. [55] used a convolutional social pool to calculate the interaction vector among vehicles and import the vector to the decoder to predict trajectories of the target vehicle. Ji et al. [56] imported all information of the target vehicle and its surrounding vehicles into the network and let the model implicitly learn the interaction among these vehicles, and the results based on the NGSIM data set show that the prediction performance is better than those of the comparative methods. The interaction-aware trajectory prediction models are closer to the actual traffic situation and can make predictions of trajectories that are following traffic rules and do not conflict with their surrounding vehicles.

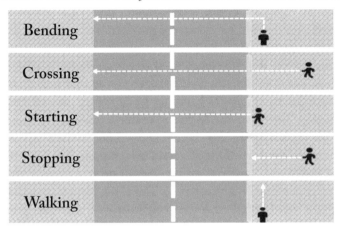

Figure 1.4: Pedestrian intention categories.

1.2 INTENTION AND TRAJECTORY PREDICTION OF THE PEDESTRIANS

Pedestrians are important traffic participants, and their movements are sometimes uncertain, which could bring great challenges to intelligent vehicle decision-making. Therefore, it is necessary to focus on pedestrian intention prediction and trajectory prediction, such that the intelligent vehicle can predict the intention and future trajectory of pedestrians, deceleration can be taken in advance to improve vehicle driving safety.

As shown in Figure 1.4, common pedestrian intentions include bending, crossing, starting, stopping, and walking. Since the pedestrian intention is sequential, it is difficult to predict pedestrian intention based on a single frame of an image. In addition, different pedestrian intentions often change from one to another, which also brings greater difficulty to the intention prediction based on the obtained images. Existing research usually considers the characteristics of pedestrian motion at multiple moments and selects one suitable algorithm model to realize the prediction of pedestrian intention. For example, Quintero et al. used pedestrian poses as the motion feature and mapped the human motion feature to the latent space through a balanced Gaussian Process Dynamical Models (GPDM), and then used a Bayesian classifier to discriminate pedestrian intentions [57]. Volz et al. used dense neural networks to determine whether pedestrians would cross the road [58]. Koehler et al. used the histogram of the oriented gradient of the contour map to describe the pedestrian motion and selected the linear support vector machine (SVM) as the intention recognition model. Since the motion contour map needs to use stereo vision information, this method has greater difficulty in obtaining data resources [59]. Moreover, Fang et al. trained three SVM classifications based on pedestrian skeleton information and predicted four pedestrian intentions including bending, crossing, starting, and stopping [60]. However, the process of training multiple SVM classifiers is more complicated. To solve the problem of SVM, Ghori et al. constructed an LSTM pedestrian intention prediction model based on the skeleton sequence and achieved the prediction of multiple pedestrian intentions in one model [61].

Nevertheless, the dynamic changes of the actual traffic environment may change pedestrian intentions. For example, when pedestrians realize that the situation is safe, their walking intention may suddenly change to bending. In this case, the skeleton feature cannot describe this change timely. Therefore, the intention prediction based on a single posture feature of the human body has certain limitations. Schulz et al. obtained the prediction probabilities of multiple intentions based on the pedestrian's head orientation; the experimental results showed that although the head orientation feature alone cannot meet the needs of multiple intention scenarios, it can predict bending in advance [62]. Therefore, head orientation is one of the important clues of intention transformation. A comprehensive consideration of head and skeleton features might help improve the accuracy of intention prediction.

At present, pedestrian trajectory prediction methods based on vehicle forward-looking views are mainly divided into two categories, namely, the prediction methods based on pedestrian motion models and the data-driven prediction methods. Commonly used pedestrian kinematics models include the constant velocity model (CV), constant acceleration model (CA), and constant position model (CP). Schneider et al. combined the CV model and Kalman filter to predict the future trajectory of pedestrians [63]. Keller et al. used interacting multiple models (IMM) to improve the accuracy of pedestrian trajectory prediction by introducing the CP model [64]. Quintero et al. used a balanced Gaussian process dynamics model (GPDM) to effectively model human motion and realize pedestrian trajectory prediction [57]. However, the above-mentioned kinematic model based on constant parameters is difficult to accurately describe the dynamic characteristics of pedestrian trajectories, which results in lower accuracy.

Lee et al. used the Recurrent Neural Network (RNN) model to adaptively learn network parameters and better deal with the challenge of dynamic changes in pedestrian motion [65]. However, due to the problem of "gradient disappearance," RNN cannot effectively model long-term sequences. The LSTM network, an improved network of RNN, can effectively alleviate this problem. Therefore, LSTM is widely used in this problem of timing relationships. For example, Hug et al. proposed a multi-modal pedestrian trajectory prediction method that combined particle filter sampling strategy and LSTM network [66]. Saleh et al. constructed multiple LSTM models for pedestrian trajectory prediction in different scenarios, and the results show that LSTM can model long-term sequences more effectively than the kinematic model [67, 68]. It can be concluded that the data-driven model based on the LSTM network can more effectively learn the complex patterns of pedestrian movement, extract abundant trajectory feature timing information, and improve the accuracy of trajectory prediction. However, the above-mentioned studies based on the LSTM method only consider the characteristics related to the historical trajectory of the pedestrian (such as historical position and speed). The general trajectory prediction models obtained by this type of method are difficult to provide accurate trajectory prediction for pedestrians with dynamic trajectories. The future trajectory of pedestrians is essentially a concrete manifestation of their walking intentions. Therefore, considering pedestrian intentions in pedestrian trajectory prediction is expected to further improve its prediction performance, which will be discussed in detail in Chapter 3.

1.3 DRIVING BEHAVIOR RECOGNITION

Recognizing the driver's activity correctly while driving is a challenging task as drivers usually have limited ranges of spatial movements, as well as narrow durations in the temporal dimension. Various activity analyses and feature extraction methods have been presented in previous research by traditional machine learning approaches and deep learning networks.

Traditional machine learning methods have the advantages of simple structures and high computational efficiency, so they are widely used in driver behavior recognition. For example, Vicente et al. used three-dimensional spatial geometric inference operations and support vector machine algorithms to detect the driver's distraction [69]. Yan et al. used the pyramid direction gradient histogram to construct driver motion characteristics and used a random forest classifier to identify shifting, calling, eating, and smoking behaviors [70]. Liang et al. use the Bayesian model to recognize the cognitive distraction of the driver in real-time [71]. Moreover, Liang et al. also use the support vector machine (SVM) to build a driver's cognitive distraction classifier [72]. Tran et al. used foot optical flow information and the hidden Markov model to construct the timing characteristics of the driver's foot motion, thereby predicting the driver's acceleration and deceleration behavior [73]. To utilize temporal information features, Craye et al. used driver's eye movement information, arm positions, head postures, and expression information as model inputs, and used the Hidden Markov Model and Adaboost classifier to recognize driver's distracted behavior [74]. Braunagel et al. [75] used an eye tracker to collect the driver's eye movement information such as the number of eye scans, gaze duration, and blink frequency and combined with the head angle to form the driver's head-eye movement model, using the SVMs to identify watching movies, watching the news, writing emails, etc. Almahasneh et al. used the singular value demarcation method to extract the driver's electroencephalography signal and trained the classifier to obtain the driver's cognitive distraction detection model [76].

However, traditional machine learning methods usually need to design algorithms such as gradient histograms, singular value demarcation methods, texture edge features, etc. to realize feature extraction. Such features generally cannot adapt to many scenes, therefore their applications could be limited. While deep learning algorithms have the advantages of strong feature learning and nonlinear representation, they also have good robustness against noise and disturbances, strong parallel computing ability, and high precision. Therefore, deep learning, which was first proposed by Professor Hinton in 1980, has gradually become the research hits, and the network could be trained through backpropagation [77]. However, deep learning failed to achieve significant developments for a long period because the computing ability of the computer was very limited and lacking large-scale datasets for the training. Until 2010, Berg et al. released a large-scale image classification dataset ImageNet, and held the ILSVRC image classification challenge [78]. In 2012, Krizhevsky et al. proposed a convolutional neural network (CNN) AlexNet, which has greatly improved the recognition accuracy on the ImageNet dataset [79]. The AlexNet network has achieved good image classification results while keeping the network concise and has become an important reference for many subsequent deep neural network studies. Since then, with the rapid improvement of computing power and the release of large datasets in various fields, deep neural networks have been applied to target detection, text recognition, action recognition, video understanding, and so on [80, 81].

Given the powerful feature extraction capabilities of deep learning, excellent model robustness, and good generalization, many studies have used deep learning methods to solve driver behavior recognition problems. For example, Okon et al. used a pre-trained CNN to implement multi-task driver behavior classification, which designed a Triple Loss function that considers the loss of positive samples, negative samples, and anchor frames at the same time to improve the recognition accuracy of the model [82]. As the continuous changes in the image background will introduce a lot of noise and affect the recognition accuracy of the model when dealing with the daily driving scene, therefore, the Gaussian mixture model is used to separate the skin color area from the image, as well as the head area and hand area. Then multiple image segmentation such as human body regions are imported into the R*CNN model, considering multi-scale and multi-region images, and more effectively classifying results of the driver behavior can be obtained [83]. Besides, Tran et al. developed a driving simulator for driver distraction recognition [84], which uses four common CNN—VGG-16, AlextNet, GoogleNet, and ResNet—to classify driver behavior, the experiments show that GoogleNet obtains the best recognition performance. Eraqi et al. used the method of ensemble learning and five parallel CNNs combined through genetic algorithms, the integrated CNN has improved accuracy compared with a single network, but the amount of model parameters is large and the computational efficiency is reduced [85].

Considering that human movement is a highly continuous process in time, some researchers started from a sequence of consecutive frames and tried to extract high-dimensional temporal features between sequences. For example, Martin et al. extracted the key point coordinates of the human skeleton from the two-dimensional image, divided the cockpit interior space into multiple areas, calculated the distance between the key point and each cockpit area, as contextual information, and finally used a multi-stream recurrent neural network Complete the recognition of the second driving task [86]. Olabiyi et al. combined camera images and vehicle motion information and used a bidirectional cyclic neural network to implement a driver behavior prediction system [87]. Peng et al. proposed a driver operation monitoring system, which uses the VGG-19 network to extract the semantic information of a single moment based on the information of vehicle trajectory and the collected images, then sends the semantic information at each moment to the LSTM network to recognize the driver operations [88]. Weyers et al.

used the recurrent neural network to classify the driver's behavior, which is based on the driver's three-dimensional key point coordinates and the segmented image of the driver's hand [89].

As a matter of fact, driver behavior recognition is a specific application of human motion recognition. In recent years, to better extract the characteristics of human motion, many related studies are committed to base on the key points of the human skeleton, which better integrate the spatial features of human motion and timing dependence. For example, Liu et al. proposed a tree-like skeleton key point traversal framework and added a new gate mechanism to the LSTM network unit so that the model can consider the feature extraction of space and time [90]. Zhang et al. proposed a Geometric-LSTM network, which converts the coordinates of the original skeleton key points into a variety of geometric feature information, such as the distance between two key points, the distance between the key point and the skeleton connection, the angle between the two skeleton connections, etc., and use the geometric feature information as the input of a three-layer LSTM network to recognize human action classification [91]. Because the simple human skeleton key point set, or its feature variants lack sufficient spatial structure expression ability, some researchers proposed to use a graph convolutional network to extract high-dimensional spatial structure features from the original skeleton key point information. For instance, Yan et al. used a graph convolutional network to realize human action recognition for the first time [92].

1.4   DRIVING STYLES

Elander [93] defines that driving style as the way that the individuals choose to drive, which includes the choice of driving speed, threshold for overtaking, expected time headway, and propensity to commit traffic violations, etc., while Berry [94] defines driving style as "the level of driving aggressiveness which includes the interaction of other traffic participants, and the driving environment." It has been found that driving style might be associated with driving safety, vehicle fuel assumption, and driving comfort, and several studies based on real driving experiments show that vehicle energy consumption is significantly influenced by driving style [95, 96]. Wu et al. [97] believed that drivers try to strike a balance between travel durations and fuel economy. Moreover, almost 56% of deadly crashes between 2003 and 2007 in the U.S. with unsafe driving behaviors were associated with aggressive driving [98]. However, if the driving style could be identified, the drivers might be likely to alter their driving behaviors, which helps to improve driving safety [99]. Even in the era of autonomous driving, research on the driving style is still vital. As we may know, autonomous driving has the potential to improve the safety and efficiency of future transportation, however, improving its acceptance for its users is an open issue and it is still a long journey to finish. Hartwich et al. [100] stated that the driver will develop a higher degree of pleasure and acceptance if the driving style of an autonomous vehicle makes a driver feel familiar and comfortable. Also, Basu et al. [101] concluded that people prefer the driving style of an autonomous vehicle is similar to their driving style. Hence, the autonomous vehicle should have different driving styles adapted to different people [102].

There have been many studies on driving style evaluation or classification over recent decades, which are based on the correlation between driving style and vehicles' performance. These evaluations for driving style are usually from subjective or objective ways.

1. *Subjective evaluation*: Subjective evaluation of driving style is mainly conducted by analyzing drivers' self-reported questionnaires. Studies from [103–106] designed questionnaires for assessing driving style, then obtained the driving style of drivers by using factor analysis. Taubman et al. [107] examined the relationship between driving style and drivers' demographics such as gender, age, driving experience, and personality by analyzing the questionnaire results. However, Basu et al. [101] found that 46–67% of drivers failed to describe their style accurately in different driving tasks, and their actual driving style was more aggressive than their self-assessment. Although subjective evaluations are significantly associated with drivers' real driving styles [108], they might not fully reflect drivers' true driving preferences and habits.

2. *Objective evaluation*: Objective evaluation realizes the driving style evaluation by analyzing the driver's operations, vehicle status, and environmental conditions. For example, Vaitkus et al. [109] presented a driving style identification method based on the signals acquired from accelerometer data, in which they used the K nearest neighbor (KNN) algorithm to classify drivers' driving styles. Murphey et al. [110] conducted a simulator study to identify the mean value of jerk on different types of roads and classified drivers' styles by comparing the jerk values of each vehicle within a specified window. However, signals in the time domain may not fully reveal their unique features of driving style, thus, spectral features from the frequency domain have been adopted. For instance, Miyajima et al. [111] used the spectral features of the accelerator pedal signal to build individual driver models. Likewise, Shi et al. [112] classified driving style based on the power spectral density of acceleration signals analysis. Chen et al. [113] applied the PCA and the DBSCAN (Density-based Spatial Clustering of Applications with Noise) to evaluating driving styles, comprehensive indicators were obtained as the main influencing factors to classify the driving style, which achieved good driving style recognition results and further motives the study of car-following driving style classification in Chapter 5.

1.5   DRIVER CHARACTERISTICS RELATED TO RISKY DRIVING BEHAVIORS

It is an undeniable fact that insufficient driving experience can lead to many driving safety problems, such as insufficient driving skills [114] and poor ability to handle driving tasks [115]. However, considering that aggressive driving is responsible for a large proportion of accidents, driving experience might not be the direct cause of risky driving behaviors, it seems to be more of a remote factor, which is mediated by certain psychological factors [116]. As for the drivers' characteristics related to risky driving behavior adopted in this short book, which includes the drivers' demographics (age and cumulative driving years), personality traits (sensation seeking), and psychological factors (risk perception). Their definitions and the relationships among them will be elaborated in brief.

1.5.1   DEMOGRAPHICS

Drivers' demographical factors are usually treated as a series of important independent or cooperative variables that affect driving behaviors in many ways. Risky driving behaviors and crash accidents are related to demographical factors such as age, gender, driving experience, and education level, with gender and age being the most significant factors [118]. Some researchers have explored relationships among different demographics and certain psychological factors, which show that the driver's demographic characteristics, such as gender and age, can significantly affect the personality traits [119, 120], which is helpful to understand the role of demographical characteristics in driving safety.

1.5.2   SENSATION SEEKING

The need for stimulation will be presumably manifested in many aspects of behaviors, including sensation, social contact, and thrill-seeking activities. The sensation-seeking scale (SSS) was developed in an early attempt to assess individual differences in optimal levels of stimulation [121], which evaluates the tendency of people to seek experiences with strong feelings and take multiple risks [122]. Sensation-seeking can be decomposed into four representative factors including *Thrill and Adventure Seeking* (*TAS*, a desire to engage in outdoor sports or other activities), *Experience Seeking* (*ES*, whose essence is "experience for its own sake."), *Disinhibition* (*DIS*, the loss of social inhibitions), and *Boredom Susceptibility* (*BS*, a dislike of repetition of experience). In terms of the impact of sensation seeking on driving safety, sensation seeking is one of the personality traits that are potentially related to risky driving behaviors. The stronger the sensation-seeking tendency, the more likely the driver is to pursue stimulation, which ultimately leads to the adoption of risky driving behaviors [123]. However, the effect of sensation-seeking on risky driving seems to be indirect. The direct effects of personality traits, especially sensation seeking, were few. Still, these personality traits have significant correlations with aberrant driving behaviors, which shows that sensation seeking is a distal but important predictor of negative driving outcomes [116]. Sensation-seeking also has a strong correlation with risk perception. However, studies focus on the effect of these two factors in the area of traffic safety are still few. Thus, this study will further explore the relationship between sensation seeking and risk perception regarding traffic safety.

1.5.3   RISK PERCEPTION

Perception of risk is a subjective judgment of the severity of a particular risk in traffic psychology, which can influence decision-based behaviors [124]. Risk perception depends on various factors such as driving training, driving experience, age, etc. Besides, risk perception is found to be negatively correlated with risky driving behaviors. Meanwhile, risk perception mediates the effect of age and gender on risky driving behaviors [116]. Moreover, risk perception does not seem to be a direct influence factor for risky driving behaviors, some scholars have discussed the relationship between the risky driving attitude and risk perception. Attitudes can be defined as "tendencies to evaluate an entity with some degree of favor or disfavor, ordinarily expressed in cognitive, affective and behavioral responses" [125], while risk-taking attitude reflects the driver's tendency toward risky driving intention, which will then affect the driver's risky driving behaviors and be regulated by risk perception.

1.5.4   MOTIVATIONS

The reason for focusing on the drivers' personality traits (e.g., sensation seeking) is that they are usually correlated with abnormal driving behaviors, which can be important influence factors of risky driving behaviors. Sensation seeking is an important personality trait that significantly affects driving behaviors, for example, impulse and sensation seeking are positively correlated with the DBQ scores [126]. However, sensation-seeking also seems not to completely affect risky driving behaviors directly, many researchers have found that personality traits also indirectly affect accident involvement through potential mediators, such as the attitude to driving safety and risk perception [127]. Risk perception and perceived benefits mediate the relationship between sensation-seeking and adolescent's risk-taking behavior [128].

However, it should be noted that gender could be another important demographical factor that may significantly influence the risky driving behaviors of the driver. It has been shown that male drivers typically report more violations [129], and engage in more risky driving behaviors than female drivers [116]. Moreover, young males are known to have an overall higher crash risk than females of the same age [130]. Besides, gender is also found to be related to personality traits, which might be an effective factor to explain the difference in risky driving behaviors between different genders. The strong influence of gender on accident propensity can be attributed in part to personality differences between male and female drivers [131]. Drivers' personality differences between different genders are also reflected in risky driving behaviors. There is a significant correlation between personality traits and abnormal driving behaviors. As a result, the influence of drivers' driving experience on their personality traits, as well as personality traits on risky driving behaviors, might be related to gender. Therefore, gender, as a moderating variable, will be adopted in this study to explore its interaction with driving experience through sensation seeking and risk perception.

Therefore, studying psychological factors related to driving experience will be helpful to understand the inhibitory relationship between driving experience and risky driving behaviors. As sensation-seeking is one of the very typical personality traits in traffic psychology, it is worthy to investigate its relationship with driving experience. While risk perception is another important psychological factor associated with driving experience. Therefore, sensation seeking and risk perception are included as mediators in the mediation model, to explore their roles in the influence of driving experience on risky driving behaviors in this chapter. In brief, we will explore the development trends of risky driving behaviors with the change of sensation seeking, risk perception, as well as the increase of driving experience between different genders, which will be discussed in Chapter 6.

CHAPTER 2

Trajectory Prediction of the Surrounding Vehicle

In this chapter, we will concentrate on predicting the future trajectories of surrounding target vehicles. The accurate prediction will help the intelligent vehicle system to make appropriate decisions and improve driving safety, efficiency, and comfort. For the convenience of description, the relevant terminology regarding vehicles used in this chapter is defined as: (i) *Host vehicle*: the vehicle is equipped with the trajectory prediction system and related sensors, which corresponds to the blue vehicle as shown in Figure 2.1a. (ii) *Target vehicle*: the vehicle that the host vehicle, shows interest which corresponds to the red vehicle, as shown in Figure 2.1a. (iii) *Surrounding vehicle*: the vehicles located around the host vehicle (except the target vehicle) correspond to the grey vehicles, as shown in Figure 2.1a.

Moreover, the dataset from the NGSIM project [132, 133] initiated by the U.S. Federal Highway Administration (FHWA) is utilized in this study. The NGSIM project contains a total of 4 data sets, namely, the US101, I-80, Lankershim Blvd, and Peachtree Street, in which we select US101 highway data to train and test for related models. As shown in Figure 2.1b, the US101 dataset selected a 2,100 ft (640 m) long road segment which has 5 lanes and 1 ramp. The raw data of the US101 was collected through a web camera, thus, we had to perform data preprocessing procedures listed below to reduce fluctuations and errors, namely:

- *Data Extraction*: extract the required features from the original US101 data set.

- *Data filtering*: the S-G filter [134] is used to filter and smooth the data (e.g., coordinates, speed).

- *Data completion*: supply the missing features.

- *Data annotation*: annotate the maneuver label of each frame in the dataset.

The following contents in this chapter are reminded as follows, Section 2.1 introduces the commonly used methodologies for trajectory prediction, which includes the kinematical-based method, the Gaussian Process method, and the proposed 3IAP model (3I: Input-Interaction-Intention, AP: Attention-Prediction). Section 2.2 focuses on the evaluation, comparison, and verification of these models, as well as a case study for the effectiveness of the proposed 3IAP model, which is derived from a real traffic accident. Finally, conclusions are drawn in Section 2.3

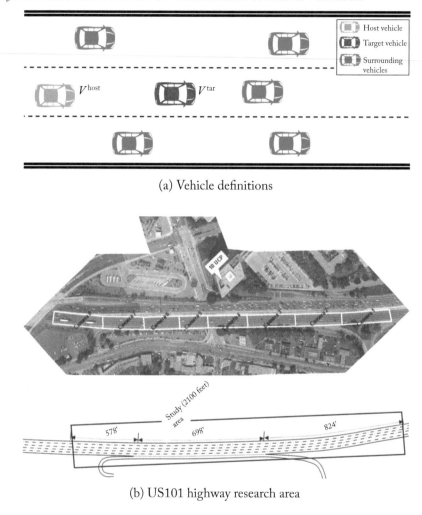

(a) Vehicle definitions

(b) US101 highway research area

Figure 2.1: Definitions and the traffic scene of the NGSIM data.

2.1   METHODOLOGIES OF THE TRAJECTORY PREDICTION

To deepen the reader's understanding of the several main-stream trajectory prediction methods as mentioned above, this section will introduce the details for each of them, respectively.

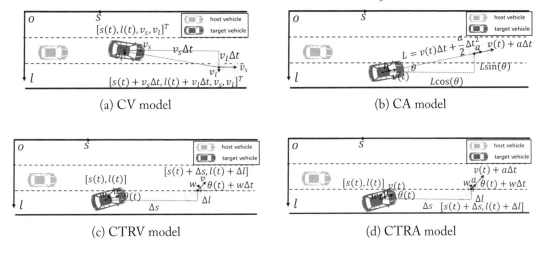

Figure 2.2: Schematic diagram of vehicle kinematic motion model.

2.1.1   THE TRAJECTORY PREDICTION MODEL BASED ON KINEMATICS

At first, several most commonly used kinematic models for trajectory prediction will be introduced, which include the constant velocity (CV) model, constant acceleration (CA) model, constant turn rate and velocity (CTRV) model, and constant turn rate and acceleration (CTRA) model.

1. **The CV Model:** the CV model assumes that the linear velocity of the vehicle is constant, and its state space X is denoted by

$$X(t) = [s(t), l(t), v_s(t), v_l(t)]^T = [s(t), l(t), v_s, v_l]^T, \qquad (2.1)$$

where $s(t), l(t)$ denote the trajectory coordinates of the target vehicle, and $v_s(t), v_l(t)$ denote the speed of the target vehicle. Since the CV model assumes that the speed is constant, and $v_s(t) = v_s, v_l(t) = v_l$. As shown in Figure 2.2a, the state transition equation is then represented by:

$$X(t + \Delta t) = \begin{bmatrix} s(t + \Delta t) \\ l(t + \Delta t) \\ v_s(t + \Delta t) \\ v_l(t + \Delta t) \end{bmatrix} = \begin{bmatrix} s(t) + \Delta t v_s \\ l(t) + \Delta t v_l \\ v_s \\ v_l \end{bmatrix}. \qquad (2.2)$$

2. **CA Model:** the CA model assumes that the acceleration of the vehicle is constant, and its state space X is denoted by:

$$X(t) = [s(t), l(t), \theta(t), v(t), a(t)]^T = [s(t), l(t), \theta, v(t), a]^T, \qquad (2.3)$$

where $s(t), l(t)$ denote the trajectory coordinates of the target vehicle, respectively; $\theta(t)$ denotes the angle between the driving direction and the S-axis (clockwise is positive) shown in Figure 2.6, and $v(t)$, $a(t)$ denote the linear velocity and acceleration of the vehicle, respectively. Since the CA model is a linear model and the acceleration is assumed to be constant where $\theta(t) = \theta, a(t) = a$. As shown in Figure 2.2b, the state transition equation is then represented by:

$$X(t + \Delta t) = \begin{bmatrix} s(t + \Delta t) \\ l(t + \Delta t) \\ \theta(t + \Delta t) \\ v(t + \Delta t) \\ a(t + \Delta t) \end{bmatrix} = \begin{bmatrix} s(t) + \left(v(t)\Delta t + \frac{a}{2}\Delta t^2\right)\cos(\theta) \\ l(t) + \left(v(t)\Delta t + \frac{a}{2}\Delta t^2\right)\sin(\theta) \\ \theta \\ v(t) + a\Delta t \\ a \end{bmatrix}. \tag{2.4}$$

3. **CTRV Model:** the CTRV model assumes that the angular velocity and linear velocity of the vehicle remain constant, and its state space X is denoted by

$$X(t) = [s(t), l(t), \theta(t), v(t), w(t)]^T = [s(t), l(t), \theta(t), v, w]^T, \tag{2.5}$$

where $s(t), l(t)$ denotes the longitudinal and transverse coordinates of the vehicle, respectively, $\theta(t)$ denotes the angle between the driving direction and the S-axis shown in Figure 2.2c, and $v(t), w(t)$ denote the linear velocity and angular velocity of the vehicle, respectively. Since the CTRV model assumes that the linear velocity and the angular velocity remain constant where $v(t) = v, w(t) = w$, the state transition equation is represented by

$$X(t + \Delta t) = \begin{bmatrix} s(t + \Delta t) \\ l(t + \Delta t) \\ \theta(t + \Delta t) \\ v(t + \Delta t) \\ w(t + \Delta t) \end{bmatrix}$$

$$= \begin{bmatrix} s(t) + \frac{v}{w}\sin(w\Delta t + \theta(t)) - \frac{v}{w}\sin(\theta(t)) \\ l(t) + \frac{v}{w}\sin(\theta(t)) - \frac{v}{w}\cos(w\Delta t + \theta(t)) \\ \theta(t) + w\Delta t \\ v \\ w \end{bmatrix}. \tag{2.6}$$

From Equation (2.6), $s(t + \Delta t)$ and $l(t + \Delta t)$ will become infinite when $w = 0$. Therefore, the case of $w = 0$ needs to be considered separately, in which the CTRV model will be equivalent to the CV model, and the corresponding state equation is represented by

$$
X\left(t + \Delta t\right) = \begin{bmatrix} s\left(t + \Delta t\right) \\ l\left(t + \Delta t\right) \\ \theta\left(t + \Delta t\right) \\ v\left(t + \Delta t\right) \\ w\left(t + \Delta t\right) \end{bmatrix} = \begin{bmatrix} s(t) + v\cos\left(\theta(t)\right)\Delta t \\ l(t) + v\sin\left(\theta(t)\right)\Delta t \\ \theta(t) \\ v \\ w \end{bmatrix}. \tag{2.7}
$$

4. **CTRA Model:** the CTRA model assumes that the angular velocity and acceleration of the vehicle remain constant, and its state space X is denoted by

$$
\begin{aligned}
X\left(t\right) &= \left[s\left(t\right), l\left(t\right), \theta\left(t\right), v\left(t\right), a\left(t\right), w\left(t\right)\right]^{T} \\
&= \left[s\left(t\right), l\left(t\right), \theta\left(t\right), v\left(t\right), a, w\right]^{T},
\end{aligned} \tag{2.8}
$$

where $s\left(t\right), l\left(t\right), \theta\left(t\right), v\left(t\right), a\left(t\right), w(t)$ denote the longitudinal and lateral coordinates, the angle between the heading direction and the S-axis as shown in Figure 2.2d, linear velocity, acceleration, and angular velocity, respectively. Since the CTRA model assumes that the acceleration and angular velocity remain constant where $a\left(t\right) = a, w\left(t\right) = w$, then the state transition equation is represented by

$$
X\left(t + \Delta t\right) = \begin{bmatrix} s\left(t + \Delta t\right) \\ l\left(t + \Delta t\right) \\ \theta\left(t + \Delta t\right) \\ v\left(t + \Delta t\right) \\ a\left(t + \Delta t\right) \\ w\left(t + \Delta t\right) \end{bmatrix} = \begin{bmatrix} s\left(t\right) + \Delta s \\ l\left(t\right) + \Delta l \\ \theta\left(t\right) + w\Delta t \\ v\left(t\right) + a\Delta t \\ a \\ w \end{bmatrix}, \tag{2.9}
$$

where

$$
\begin{aligned}
\Delta s = \frac{1}{w^2} &\left[(v(t)w + aw\Delta t)\sin\left(\theta(t) + w\Delta t\right)\right. \\
&\left. + a\cos\left(\theta(t) + w\Delta t\right) - v(t)w\sin\left(\theta(t)\right) - a\cos\left(\theta(t)\right)\right]
\end{aligned} \tag{2.10}
$$

$$
\begin{aligned}
\Delta l = \frac{1}{w^2} &\left[(-v(t)w - aw\Delta t)\cos\left(\theta(t) + w\Delta t\right)\right. \\
&\left. + a\sin\left(\theta(t) + w\Delta t\right) + v(t)w\cos\left(\theta(t)\right) - a\sin\left(\theta(t)\right)\right].
\end{aligned} \tag{2.11}
$$

Again, when $w = 0$, Δs, and Δl will become infinite. Therefore, the case of $w = 0$ needs to be considered separately, in which the CTRA model will be equivalent to the CA model, then

$$
\Delta s = \left(v(t)\Delta t + \frac{1}{2}a\Delta t^2\right)\cos\left(\theta(t)\right) \tag{2.12}
$$

Table 2.1: Test results of each kinematic model

Kinematic Models	Longitudinal APE (m)	Lateral APE (m)
CV	0.4602	2.7800
CA	0.5686	6.3008
CTRV	0.4917	2.8011
CTRA	0.6434	6.3263

$$\Delta l = \left(v(t)\Delta t + \frac{1}{2}a\Delta t^2 \right) \sin\left(\theta(t)\right). \tag{2.13}$$

Different from the machine-learning-based methods, the trajectory prediction method based on the kinematic models does not require training, while it utilizes relevant parameters of the target vehicle measured by the host vehicle's equipped sensors, and outputs the predicted trajectory directly. More specifically, if the sensors of the host vehicle at the current time t have detected the target vehicle's state $X(t)$, then we can set a small update time as Δt within the prediction horizon, and feed $X(t)$ and Δt into the CV/CA/CTRV/CTRA model, respectively, then the predicted trajectories in the future for each model could be obtained accordingly.

Based on the NGSIM-US101 dataset, we perform a test for each of the kinematic models and calculate the corresponding average prediction errors of the longitudinal and lateral direction, which the test results are shown in Table 2.1. According to Table 2.1, the APE (Average Prediction Error) in the longitudinal and lateral direction of the CV model in the test set are 0.4602 m and 2.7800 m, respectively, which is the best among them. Therefore, we select the CV model as the target model. One reason is that the curvatures of the test course are quite small and the vehicles on the road would not accelerate or decelerate frequently, however, please keep in mind that the choice of the kinematic model might vary based on the specific data. For example, if the data were mostly collected on large curved roads, intuitively, the CRTV or CTRA model could be better choices than the CV model.

2.1.2   THE TRAJECTORY PREDICTION MODEL BASED ON THE GAUSSIAN PROCESS

The Gaussian process (GP) is a kind of stochastic process, and each moment corresponds to a random variable that obeys Gaussian distribution. A Gaussian process $f(t)$ can be determined jointly by the mean function $m(t)$ and the kernel function $K(t_i, t_j)$, namely

$$f(t) \sim GP(m, K). \tag{2.14}$$

The linear combination of any finite number of random variables in the Gaussian process still obeys the Gaussian distribution, that is

$$[f(t_1), f(t_2), \ldots, f(t_n)] \sim N(\boldsymbol{\mu}, \boldsymbol{\Sigma}), \tag{2.15}$$

where $\boldsymbol{\mu} = [\mu(t_1), \ldots, \mu(t_n)]$ denotes the mean vector, which is determined by the mean function $m(t)$ of the Gaussian process, namely,

$$\mu(t_i) = m(t_i), \ i = 1, 2, \ldots, n, \tag{2.16}$$

while $\boldsymbol{\Sigma} \in \mathbb{R}^{n \times n}$ denotes the covariance matrix, which is determined by the kernel function $K(t_i, t_j)$ of the Gaussian process, namely,

$$\Sigma_{ij} = K(t_i, t_j). \tag{2.17}$$

(1) *Mean Function*:   the mean function $m(t)$ describes the variable's trend over time and the overall position corresponding to each moment. The following are several common mean functions that can be selected according to actual conditions.

- Constant mean function

$$m(t) = c, \tag{2.18}$$

  where c denotes a constant.

- Linear mean function

$$m(t) = a * t + b; \tag{2.19}$$

  where a and b are constant.

- Polynomial mean function

$$m(t) = \sum_{i=0}^{p} a_i t^i, \tag{2.20}$$

  where a_i denotes a constant coefficient.

(2) *Kernel Function*:   the properties of the Gaussian process are closely related to its kernel function $K(t_i, t_j)$, which describes the relationship between different instants t_i and t_j. The following are the commonly used kernel functions that can be selected according to the actual situation.

- Linear kernel function

$$K(t_i, t_j) = \sigma_f^2 t_i t_j. \tag{2.21}$$

- The radial basis kernel function

$$K(t_i, t_j) = \sigma_f^2 \exp\left(-\frac{(t_i - t_j)^2}{2l^2}\right). \tag{2.22}$$

- Exponential kernel function

$$K\left(t_i, t_j\right) = \sigma_f^2 \exp\left(-\frac{|t_i - t_j|}{l}\right). \tag{2.23}$$

- Periodic kernel function

$$K\left(t_i, t_j\right) = \sigma_f^2 \exp\left(\sigma_l^2 \sin^2\left(t_i - t_j\right)\right). \tag{2.24}$$

- Brown kernel function

$$K\left(t_i, t_j\right) = \sigma_f^2 \min\left(t_i, t_j\right). \tag{2.25}$$

As mentioned before, the linear combination of variables at a different time of the Gaussian process is still following Gaussian distribution and the kernel function describes the relationship between different time steps, the Gaussian process is widely used for regression analysis, namely, the Gaussian Process Regression (GPR).

Training for the GPR: We first extract a set of trajectories with different driving maneuvers (LLC, RLC, and LK), and each trajectory contains $n + m$ points (n historical trajectory points, m predicted trajectory points). Assuming that the longitudinal and lateral coordinates are independent, as shown in Table 2.2, the longitudinal and lateral trajectory sets corresponding to each maneuver can be extracted.

We construct the corresponding GPs for each trajectory set, as shown in Table 2.2. Considering the shape of the vehicle trajectory, the fifth-degree polynomial function is selected as the mean function of GPs for l_{LLC} and l_{RLC} trajectory sets, that is

$$m(t) = a_5 t^5 + a_4 t^4 + a_3 t^3 + a_2 t^2 + a_1 t^1 + a_0, \tag{2.26}$$

where a_i, $i = 0\ldots5$ denotes the polynomial coefficient of (2.26).

Also, select the linear function as the mean function of GPs for $s_{LLC}, s_{RLC}, s_{LK}, l_{LK}$ trajectory sets, namely,

$$m(t) = a * t + b, \tag{2.27}$$

where a, b denote the linear coefficient of (2.27), respectively. Considering the smoothness of the vehicle trajectory, we select the radial basis function as the kernel function of the GPs, which is

$$K\left(t_i, t_j\right) = \sigma_f^2 \exp\left(-\left(t_i - t_j\right)^2 / \left(2l^2\right)\right), \tag{2.28}$$

where σ_f and l denote the hyper-parameters of (2.28), respectively. Obviously, there are a total of 32 parameters that need to be estimated, in which those relevant parameters can be learned by the Maximum Likelihood Estimation (MLE) method based on the known historical data.

Table 2.2: Trajectory set of different maneuvers

Trajectory Set	Description
$\{S_{LLC}\}$	Longitudinal coordinate trajectory set of the LLC
$\{l_{LLC}\}$	Lateral coordinate trajectory set of the LLC
$\{S_{RLC}\}$	Longitudinal coordinate trajectory set of the RLC
$\{l_{RLC}\}$	Lateral coordinate trajectory set of the RLC
$\{S_{LK}\}$	Longitudinal coordinate trajectory set of the LK
$\{l_{LK}\}$	Lateral coordinate trajectory set of the LK

Trajectory Prediction by the GPR: According to the properties of the GPs, the $n + m$ trajectory coordinates obey the joint Gaussian distribution, namely:

$$[s_{t-m+1}, \ldots, s_t, \ldots, s_{t+n}] \sim N\left(\boldsymbol{\mu}^s, \boldsymbol{\Sigma}^s\right); \tag{2.29}$$

$$[l_{t-m+1}, \ldots, l_t, \ldots, l_{t+n}] \sim N\left(\boldsymbol{\mu}^l, \boldsymbol{\Sigma}^l\right). \tag{2.30}$$

Given n historical trajectory points, the probability distribution of trajectory points at m time in the future can be derived. Let us take the longitudinal coordinate l as an example, the $n + m$ longitudinal coordinates can be divided into two groups: historically known coordinates $l_h = [l_{t-n+1}, \ldots, l_t]$, future predicted coordinates $l_f = [l_{t+1}, \ldots, l_{t+m}]$, such that

$$l = \begin{bmatrix} l_h \\ l_f \end{bmatrix}. \tag{2.31}$$

Then the trajectory prediction problem of the longitudinal coordinate is equivalent to calculating the conditional probability distribution $P(l_f|l_h)$. Also, divide the mean vector and covariance matrix:

$$\boldsymbol{\mu}^l = \begin{bmatrix} \boldsymbol{\mu}_h^l & \boldsymbol{\mu}_f^l \end{bmatrix}^T \tag{2.32}$$

$$\boldsymbol{\Sigma}^l = \begin{bmatrix} \boldsymbol{\Sigma}_{hh}^l & \boldsymbol{\Sigma}_{hf}^l \\ \boldsymbol{\Sigma}_{fh}^l & \boldsymbol{\Sigma}_{ff}^l \end{bmatrix}, \tag{2.33}$$

where $\boldsymbol{\Sigma}_{hf}^l = \boldsymbol{\Sigma}_{fh}^{l\ T}$, and $\boldsymbol{\Sigma}_{hh}^l, \boldsymbol{\Sigma}_{ff}^l$ are symmetric matrix.

If both l_f and l_h follow the Gaussian distribution, then the conditional probability $P(l_f|l_h)$ is still Gaussian, that is $l_f|l_h \sim N(\boldsymbol{\mu}_{f|h}^l, \boldsymbol{\Sigma}_{f|h}^l)$, such that $\boldsymbol{\mu}_{f|h}^l, \boldsymbol{\Sigma}_{f|h}^l$ can be calculated by

$$\boldsymbol{\mu}_{f|h}^l = \boldsymbol{\mu}_f^l + \boldsymbol{\Sigma}_{fh}^l \left(\boldsymbol{\Sigma}_{hh}^l\right)^{-1} \left(l_h - \boldsymbol{\mu}_h^l\right) \tag{2.34}$$

$$\Sigma_{f|h}^{l} = \Sigma_{ff}^{l} - \Sigma_{fh}^{l} \left(\Sigma_{hh}^{l}\right)^{-1} \Sigma_{hf}^{l}. \tag{2.35}$$

Finally, the conditional probability $P\left(l_f|l_h\right)$ can be represented by

$$P\left(l_f|l_h\right) = \frac{1}{\sqrt{(2\pi)^n \left|\Sigma_{f|h}^{l}\right|}} \exp\left(-\frac{1}{2}\left(l - \mu_{f|h}^{l}\right)^T \left(\Sigma_{f|h}^{l}\right)^{-1} \left(l - \mu_{f|h}^{l}\right)\right). \tag{2.36}$$

Similarly, the conditional probability $P\left(s_f|s_h\right)$ corresponding to the longitudinal coordinate is denoted by

$$P\left(s_f|s_h\right) = \frac{1}{\sqrt{(2\pi)^n \left|\Sigma_{f|h}^{s}\right|}} \exp\left(-\frac{1}{2}\left(s - \mu_{f|h}^{s}\right)^T \left(\Sigma_{f|h}^{s}\right)^{-1} \left(s - \mu_{f|h}^{s}\right)\right). \tag{2.37}$$

Thus, predicted trajectories $[l_{t+1}, \ldots, l_{t+m}]$ and $[s_{t+1}, \ldots, s_{t+m}]$ of the target vehicle in the future can be obtained.

2.1.3   THE TRAJECTORY PREDICTION MODEL BASED ON THE 3IAP

Compared with the physics-based methods and the maneuver-based methods, the interaction-aware trajectory prediction models take the interaction among surrounding vehicles into consideration, which is more complete and has a better prediction effect. As shown in Figure 2.3, we propose a so-called 3IAP trajectory prediction model (3I: Input-Interaction-Intention, AP: Attention-Prediction), which is a kind of implicit modeling interaction based on the deep neural network. It includes five key modules including the input module, interaction module, intention recognition module, attention module, and trajectory prediction module. Next, we will briefly introduce the configuration of each module.

Input Module: The input module extracts the following information: coordinates of the target vehicle X^{tar}, coordinates of surrounding vehicles X^{s_i} ($i = 1, 2, \ldots, k$, k is the number of surrounding vehicles) and the Space Location Grid (SLG). The coordinates of the target vehicle X^{tar} and surrounding vehicles X^{s_i} can be obtained directly from the data set. And the SLG can be extracted in the following way:

- Regard the target vehicle as to the reference point, and divide the road into grids, the size of each grid is $L \times W$, where W denotes the width of the lane, and L denotes the average length of the vehicle.

- Take 20 grids in front of the target vehicle and 10 grids behind the target vehicle to construct the SLG, which contains 31×3 grids, such that the target vehicle is in row 21, column 2 of the SLG.

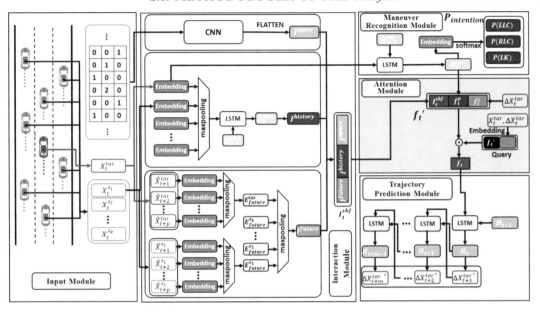

Figure 2.3: Structure diagram of the 3IAP model.

- Fill the position information into the SLG. The grid occupied by the target vehicle and surrounding vehicles is replaced by a dummy variable 2 and 1, respectively, while the empty grids are filled with 0. When the target vehicle is located in the leftmost (or rightmost) lane, as shown in Figure 2.4, since there is no lane on its left (or right) to travel, we fill the corresponding grid with -1 in this situation.

Interaction Module: The interaction module contains three components of interaction features: spatial interaction features (I^{space}), historical interaction features ($I^{history}$), and the future inter-active features (I^{future}). More specifically, the spatial interaction features describe the relative spatial relationship among the target vehicle and its surrounding vehicles at the instant t. We use a CNN to extract the features of the SLG, and then flatten the feature map to obtain I^{space}, that is:

$$I^{space} = \text{flatten}\left(\text{CNN}\left(SLG\right)\right). \tag{2.38}$$

At each historical instant t, we first embed the location of the target vehicle and its surrounding vehicles and extract the interaction information by the maximum pooling operation, then we import the interaction information into an independent LSTM network and obtain the hidden variable H_t^{hist}, which contain historical interaction information. Finally, we embed H_t^{hist} to extract the historical interaction features $I^{history}$ among the target vehicle and its surrounding

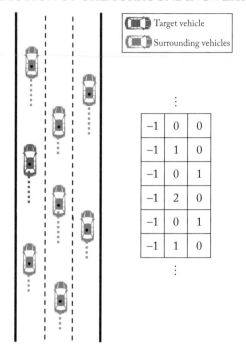

Figure 2.4: Schematic diagram of the target vehicle in the leftmost lane.

vehicles, namely,

$$E_t^{object} = \text{Embedding}\left(X_t^{object}\right), \qquad object = tar, s_1, \ldots, s_k; \qquad (2.39)$$

$$H_t^{hist} = LSTM\left(\text{maxpooling}\left(\left[E_t^{tar}, E_t^{s_1}, \ldots, E_t^{k}\right]\right), H_{t-1}^{hist}\right), \qquad i = 1, 2, \ldots, k; \quad (2.40)$$

$$I^{history} = \text{Embedding}\left(H_t^{s}\right). \qquad (2.41)$$

Moreover, the 3IAP uses the *CV* model to predict the coarse-grained trajectories of the target vehicle and its surrounding vehicles for the future p-steps firstly, and then calculate the interaction between these future trajectories to obtain the future interaction features, namely,

$$\left[\widehat{X}_{t+1}^{tar}, \widehat{X}_{t+2}^{tar}, \ldots, \widehat{X}_{t+p}^{tar}\right] = CV\left(X_t^{tar}\right); \qquad (2.42)$$

$$\left[\widehat{X}_{t+1}^{s_i}, \widehat{X}_{t+2}^{s_i}, \ldots, \widehat{X}_{t+p}^{s_i}\right] = CV\left(X_t^{s_i}\right); \qquad (2.43)$$

$$E_{future}^{tar} = \text{maxpooling}\left(\text{Embedding}\left(\left[\widehat{X}_{t+1}^{tar}, \widehat{X}_{t+2}^{tar}, \ldots, \widehat{X}_{t+p}^{tar}\right]\right)\right); \qquad (2.44)$$

$$E_{future}^{s_i} = \text{maxpooling}\left(\text{Embedding}\left(\left[\widehat{X}_{t+1}^{s_i}, \widehat{X}_{t+2}^{s_i}, \ldots, \widehat{X}_{t+p}^{s_i}\right]\right)\right); \qquad (2.45)$$

$$I^{future} = \text{maxpooling} \left(\left[E^{tar}_{future}, E^{s_1}_{future}, \ldots, E^{s_k}_{future} \right] \right). \tag{2.46}$$

Finally, I^{space}_t, $I^{history}_t$, and I^{future}_t are concatenated together to get the interaction feature I^{shf}_t, namely:

$$I^{shf}_t = \text{concat} \left(\left[I^{space}_t, I^{history}_t, I^{future}_t \right] \right). \tag{2.47}$$

Intention Recognition Module: The intention recognition module is used to recognize the driving maneuver of the target vehicle, to improve the prediction performance. The output of this module contains two parts, which includes the maneuver feature f^P_t and the maneuver probability $P_{intention}$. We use an independent LSTM to encode X^{tar}_t, and take the LSTM unit's hidden state H^{tar}_t at the current instant as the intention feature f^P_t, which will be directly imported to the downstream module. And we also decode H^{tar}_t to get the probability ($P_{intention}$) of different maneuvers by the softmax function, which will be further used in the calculation of the final loss.

$$f^P_t = H^{tar}_t = LSTM \left(\text{Embedding} \left(X^{tar}_t \right), H^{tar}_{t-1} \right); \tag{2.48}$$

$$P_{intention} = \text{softmax} \left(\text{Embedding} \left(H^{tar}_t \right) \right). \tag{2.49}$$

Attention Module: The interaction features I^{shf}_t and the maneuver features f^P_t, the motion feature f^Δ_t could be calculated by

$$f^\Delta_t = \text{Embedding} \left(\Delta X^{tar}_t \right). \tag{2.50}$$

The features $f'_t = \text{Concat} \left(\left[I^{shf}_t, f^P_t, f^\Delta_t \right] \right)$ can be directly imported to the LSTM decoder and export the predicted trajectories of the target vehicle. However, the importance of the above three features for predicting the trajectory is different, and the importance of different sub-features within one feature is also different. Thus, we calculate the importance of each feature via the attention mechanism and weigh it according to the importance, eventually, get the final features f_t. In addition, the query Q, key K, and value V of the attention module could be denoted by

$$Q = \text{Concat} \left(\left[\Delta X^{tar}_t, X^{tar}_t, f'_t \right] \right); \tag{2.51}$$

$$K = f'_t; \tag{2.52}$$

$$V = f'_t. \tag{2.53}$$

The additive attention model is adopted in this study, such that,

$$f_t = \text{mul} \left(V, \text{softmax} \left[W^T_S \tanch \left(W_Q Q + W_K K \right) \right] \right), \tag{2.54}$$

where mul() denotes the matrix multiplication.

Table 2.3: Parameter settings of the 3IAP model

Parameters	Description	Value
n	Length of historical track sequence	15
m	Length of predicted track sequence	25
d_{model}	Model dimensions	32
p	Coarse-grained prediction of the number of track points	15
l_r	Learning rate	1e-4
λ	Weight between longitudinal and lateral trajectory loss	20
γ	Weight between initial trajectory loss and maneuver loss	100
batch size	Batch size	128
activation	Activation function	tanh function
optimizer	Optimizer	Adam optimizer
loss function	Loss function	Custom

Trajectory Prediction Module: The features f_t are imported into the LSTM decoder, and export the predicted displacements of the target vehicle for the future m-steps, that is

$$\left[\Delta X_{t+1}^{tar\ast}, \Delta X_{t+2}^{tar\ast}, \ldots, \Delta X_{t+m}^{tar\ast}\right] = LSTMDecoder\left(f_t\right). \tag{2.55}$$

Thus, the predicted trajectories can be calculated:

$$X_{t+1}^{tar\ast} = X_t^{tar} + \Delta X_{t+1}^{tar\ast};$$

$$X_{t+2}^{tar\ast} = X_{t+1}^{tar\ast} + \Delta X_{t+2}^{tar\ast};$$

$$\vdots$$

$$X_{t+m}^{tar\ast} = X_{t+m-1}^{tar\ast} + \Delta X_{t+m}^{tar\ast}. \tag{2.56}$$

According to experience and proper parameter adjustments, we set the hyper-parameters shown in Table 2.3. The loss function in Table 2.3 is a custom loss, which consists of three parts, including the MSE loss of longitudinal trajectories, MSE loss of lateral trajectories, and cross-entropy loss for maneuver recognition.

$$MSE_s = \frac{1}{N} \sum_{i=1}^{N} \frac{1}{m} \sum_{t=1}^{m} \left[\left(s_{i,t}^{\ast} - s_{i,t}\right)^2\right]; \tag{2.57}$$

$$MSE_l = \frac{1}{N} \sum_{i=1}^{N} \frac{1}{m} \sum_{t=1}^{m} \left[\left(l_{i,t}^{\ast} - l_{i,t}\right)^2\right]; \tag{2.58}$$

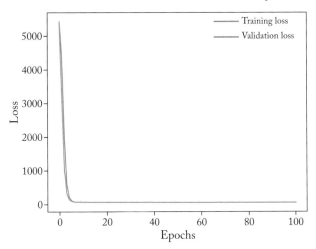

Figure 2.5: Loss change curve of 3IAP model.

$$CE = -\frac{1}{N} \sum_{i=1}^{N} \sum_{k=1}^{K} p_{ik} \log(q_{ik});$$ (2.59)

$$Loss = (RMSE_s + \lambda * RMSE_l) + \gamma CE,$$ (2.60)

where N denotes the number of sample sequences; m denotes the number of predicted trajectory points; $s_{i,t}^*, l_{i,t}^*$ denotes the lateral and longitudinal prediction coordinates of the i-th sample by the 3IAP model at the prediction instant t, respectively, $s_{i,t}, l_{i,t}$ denotes the true lateral and longitudinal coordinates corresponding to the i-th sample at the prediction instant t, respectively, K denotes the number of maneuvers, p_{ik} denotes the ground-truth label in one-hot form, and q_{ik} denotes the probability that the i-th sample predicted by the model belongs to the k-th maneuver; λ denotes the weight between the longitudinal trajectory MSE loss and the lateral trajectory MSE loss, and γ denotes the weight between the trajectory loss and the maneuver loss.

We use the PyTorch framework to construct the 3IAP model as shown in Figure 2.3 and set the parameters as shown in Table 2.3. The training strategy of the 3IAP model is: (1) we train the maneuver branch network for 10 epochs, which means only use the maneuver loss to update the parameters; (2) train the entire network for 100 epochs, which means update the parameters of the entire network with the custom loss shown in formulas (2.57)–(2.60). During the training process, the weight λ is a fixed value, and the weight γ decreases with the epoch increasing, and finally $\gamma = 0$. Finally, the curve of training loss and validation loss during training is shown in Figure 2.5.

2.2   EXPERIMENTS AND RESULTS

Based on the NGSIM-US101 data set, the aforementioned related trajectory prediction models are tested. To comprehensively compare the performance of the above methods, the following indicators are used for evaluation:

- *Average Displacement Error (ADE)*: it refers to the predicted average position error for the m prediction time steps, namely

$$ADE = \frac{1}{N} \sum_{i=1}^{N} \frac{1}{m} \sum_{t=1}^{m} E_{i.t}, \qquad (2.61)$$

where N denotes the number of samples, m is the number of predicted trajectory points, $E_{i.t}$ is the position error of the t-th point predicted by the i-th sample, and the same below.

- *Final Displacement Error (FDE)*: it refers to the predicted average position error of the last trajectory point, namely

$$FDE = \frac{1}{N} \sum_{i=1}^{N} E_{i,m}. \qquad (2.62)$$

- *Short-term Displacement Error (SDE)*: it refers to the predicted short-term (the first third of the points) average position error for the m prediction time steps, namely

$$SDE = \frac{1}{N} \sum_{i=1}^{N} \frac{3}{m} \sum_{t=1}^{m/3} E_{i.t}. \qquad (2.63)$$

- *Middle-term Displacement Error (MDE)*: it refers to the predicted average position error in the mid-term (middle third of the trajectory points) for the m prediction time steps, namely

$$MDE = \frac{1}{N} \sum_{i=1}^{N} \frac{3}{m} \sum_{t=m/3}^{2m/3} E_{i.t}. \qquad (2.64)$$

- *Long-term Displacement Error (LDE)*: it refers to the predicted long-term (the last third of the trajectory points) average position error for the m prediction time steps, namely

$$LDE = \frac{1}{N} \sum_{i=1}^{N} \frac{3}{m} \sum_{t=2m/3}^{m} E_{i.t} \qquad (2.65)$$

Table 2.4: Performance indicators of each trajectory prediction model (m)

Model		ADE(m)	FDE(m)	SDE(m)	MDE(m)	LDE(m)
CV	Longitudinal	0.4068	0.8350	0.1227	0.4054	0.6920
	Lateral	2.1244	5.3479	0.4406	1.8345	4.0338
GP	Longitudinal	0.3071	0.5943	0.1107	0.3113	0.5003
	Lateral	2.0383	5.0673	0.4452	1.7647	3.8441
3IAP	Longitudinal	0.1624	0.3045	0.0659	0.16400	0.2576
	Lateral	1.4512	4.0443	0.2777	1.1498	2.8591

Table 2.5: 3IAP model ablation experiment results

Model	Interaction Module	Intention Recognition Module	Attention Module	Longitudinal ADE(m)	Lateral ADE(m)
Baseline	✗	✗	✗	0.2564	1.8240
3IAP-V1	✓	✗	✗	0.2055	1.6535
3IAP-V2	✓	✓	✗	0.1702	1.5153
3IAP	✓	✓	✓	0.1624	1.4512

The indicators of trajectory prediction models described above are calculated, and the results are shown in Table 2.4. Based on the results, we can observe that the 3IAP model performs better than the CV-based and GP-based models respecting the ADE, FDE, SDE, MDE, and the LDE of the longitudinal and lateral direction.

2.2.1   ABLATION EXPERIMENTS

To verify the effectiveness of each module of the 3IAP, ablation experiments are further conducted, and the results are shown in Table 2.5. The baseline is the basic LSTM model, where the ADE in the longitudinal and lateral direction is 0.2564 m and 1.8240 m, respectively. The 3IAP-V1 adds an interaction module based on the baseline where the ADEs in the longitudinal and lateral direction are 0.2055 m and 1.6535 m, respectively, which are 0.0509 m and 0.1705 m lower than the baseline. The 3IAP-V2 adds an intention recognition module based on the 3IAP-V1 where the ADEs in the longitudinal and lateral directions are 0.1702 m and 1.5153 m, respectively, which are reduced by 0.0353 m and 0.1382 m, respectively, compared with the 3IAP-V1. The 3IAP model adds an attention module based on 3IAP-V2 where the ADEs in the longitudinal and lateral directions are 0.1624 m and 1.4512 m, respectively, which are 0.0078 m and 0.064 m1 lower than the 3IAP-V2. Therefore, each module in the 3IAP model can reduce the prediction error of the vehicle's future trajectory to a certain extent.

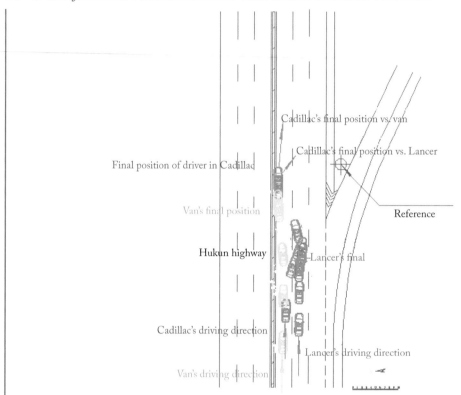

Figure 2.6: Illustration of an accident that happened on the Hu-Kun Highway in 2018.

2.2.2   CASE STUDY

We conduct a verification experiment related to a real car accident to show the effectiveness and feasibility of the proposed method in this section. As shown in Figure 2.6, a collision happened between the van's front side and the Cadillac XT5's rear side on the west side of China Hu-Kun Highway in 2018, when the Cadillac XT5 made a right lane change maneuver and collided with the Lancer. We intend to show that if the Lancer is equipped with the proposed 3IAP trajectory prediction system in this chapter, it might be able to predict the future trajectory of Cadillac XT5, thus, avoid the collision.

   According to the accident report, the Cadillac's width is 1.90 m, and the speed before the collision was 58 km/h, while the Lancer's width is 1.76 m, and the speed at the collision was 83 km/h. Since no further information was given in this report, however, we could make the following reasonable assumptions to reconstruct the accident scene: (i) the lane width is 3.5 m; (ii) before the collision, the Cadillac was moving in the lane centerline at a constant speed of

Table 2.6: Locations of Cadillac and Lancer 10 s before the collision moment

Time Before Collision (TBC)	Cadillac XT5 Location (m)		Lancer Location (m)	
	l	s	l	s
−1s	−0.4	−16.1	0	−21.3
−2s	−0.9	−32.2	0	−42.6
−3s	−1.7	−48.3	0	−63.9
−4s	−2.5	−64.4	0	−85.2
−5s	−3.1	−80.5	0	−106.5
−6s	−3.5	−96.6	0	−127.8
−7s	−3.5	−112.7	0	−149.1
−8s	−3.5	−128.8	0	−170.4
−9s	−3.5	144.9	0	−191.7
−10s	−3.5	−161.0	0	−213.0

58 km/h (16.1m/s); and (iii) before the collision, the Lancer was moving in the lane centerline at a constant speed of 83 km/h (21.3 m/s).

Such that, the locations with 10 s before the collision are shown in Table 2.6, which would be imported into the proposed 3IAP model, and the obtained simulation results of the maneuver recognition regarding the target vehicle Cadillac XT5 are depicted in Figure 2.7. When the TBC $= -5$ s, the maneuver recognition module of 3IAP tells that the Cadillac XT5 is going to perform an RLC maneuver, and the predicted trajectory of the Cadillac is generated by the 3IAP is shown in Figure 2.8. If the Lancer begins to brake with an emergency deceleration of $0.8g$ ($1g = 9.8$m/s^2), the braking-to-stop time is:

$$t_{break} = \frac{21.3 \text{ m/s} - 0 \text{ m/s}}{0.8 \times 9.8 \text{ m/s}^2} = 2.7 \text{ s}.$$

And the traveled station during this time is:

$$S_{break} = 21.3 \times 2.7 - \frac{1}{2} \times 9.8 \times 2.7^2 = 21.8 \text{ m}.$$

As shown in Figure 2.8, the Cadillac and Lancer could avoid collisions.

2.3   SUMMARY

This chapter introduced three trajectory prediction methods including the kinematical-based, the maneuver-based, and the interaction-aware-based.

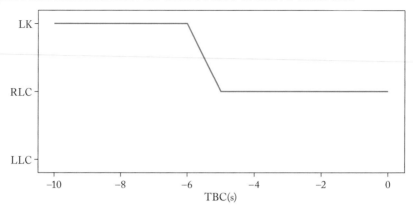

Figure 2.7: Maneuver recognition results obtained from the 3IAP.

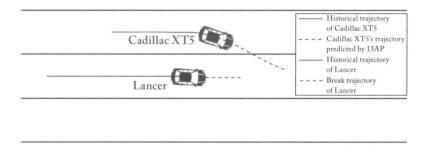

Figure 2.8: Scene diagram of the prediction trajectories.

1. The kinematical-based trajectory prediction method is simple and has the advantage of good efficiency, but it has certain limitations which only take the low-level attributes (kinematical features) of the vehicle into consideration.

2. The maneuver-based trajectory prediction method considers the influence of the vehicle's driving maneuver and has better prediction performance in the long-term predict period. However, this method does not consider the interaction between the target vehicle and its surrounding environment. In this chapter, we use the Gaussian Process model to output the probability distribution of future predicted trajectories.

3. While the interaction-aware trajectory prediction method is closer to the actual traffic situation, which can make a more accurate prediction about the target vehicle without any conflict with its surrounding vehicles. The 3IAP model exports the predicted trajectories of the target vehicle through the input module, the interaction module, the maneuver recognition module, the attention module, and the trajectory prediction module.

Based on the NGSIM-US101 data set, a comparative study on these three methods has been conducted. The results show that the lateral ADE and FDE errors of the 3IAP trajectory prediction model are 0.1624 m and 0.3045 m, respectively, and the longitudinal ADE and FDE errors are 1.4512 m and 4.0443 m, respectively, which are better than methods based on kinematics and Gaussian Process. Through ablation experiments on the 3IAP model, the interaction module, the maneuver recognition module, and the attention module can reduce the longitudinal and lateral ADE errors, such that, to show the effectiveness of each module. Finally, the proposed 3IAP model is verified by a real accident case that happened in China, we reconstructed the accident process and import relevant data into the 3IAP model. The simulation results show that the 3IAP model can recognize Cadillac's right-lane-changing maneuver by 5 s in advance, if the Lancer performs emergency braking at this time, the collision could be avoided.

CHAPTER 3

Predictions of the Intention and Future Trajectory of the Pedestrian

In this chapter, we will focus on pedestrian intention and its trajectory prediction. Pedestrians are important traffic participants, and the dynamic changes of their walking intentions and motion trajectories bring major challenges to intelligent vehicle decision-making and planning. For example, pedestrians may decide to cross the road when they were waiting on the side of the road, appearing in any position of the vehicle. Therefore, research on intention and trajectory prediction of pedestrians is of great significance for reducing the probability of collisions between vehicles and pedestrians, improving vehicle safety, and promoting the development of intelligent vehicle technology. Most of the previous studies only use single human posture features such as skeletons to predict pedestrian intentions. Considering the head orientation feature is one of the important clues of intention projection of the pedestrian, merging it with pedestrian skeleton features is expected to improve the accuracy of pedestrian intention prediction. The introduction of intention features can enhance the learning of pedestrian dynamic changes by the trajectory prediction network and improve the accuracy of pedestrian trajectory prediction. The main contents of this chapter will include the following.

1. Pedestrian intention prediction based on multiple features of skeleton and head orientation: the human pose estimation algorithm and the head orientation estimation algorithm are used to obtain the features of pedestrian skeleton and head orientation, respectively, and these two parts of features are merged to strengthen the expression of pedestrian movement changes. Based on the features obtained by the fusion, the LSTM intention prediction model is established to improve the performance of intention prediction.

2. Hierarchical pedestrian trajectory prediction fused with pedestrian intention: the intention prediction result is obtained through the pedestrian intention prediction network, and the prediction intention is merged with the historical trajectory. In addition, the attention mechanism is introduced in the encoder-decoder of the LSTM network to enhance timing feature extraction, improving the performance of trajectory prediction.

| (a) Bending | (b) Crossing | (c) Starting | (d) Stopping |

Figure 3.1: Examples of the pedestrian movement.

Table 3.1: Data distribution of train set and test set

Categories	Training Set	Test Set	Total
Bending	121	78	199
Crossing	97	72	169
Starting	48	43	91
Stopping	41	26	67
Walking	66	33	99
Total	373	252	625

3.1   DATA PREPARATION

The commonly used pedestrian-related datasets include Daimler, INRIA, MOT, and Caltech, etc. However, considering that none of the datasets except the Daimler contain continuous pedestrian motion images and vehicle trajectory coordinates, therefore, this study is based on the Daimler dataset. The Daimler dataset consists of a training set and a test set. There are a total of 68 long pedestrian motion sequences with a sampling frequency of 16 Hz. The dataset also contains the original image, the position coordinates of the pedestrian relative to the vehicle, as well as the speed of the vehicle. As shown in Figure 3.1, pedestrian movement in the dataset includes bending, crossing, starting, and stopping. As for straight walking, it can be obtained by expanding the sequence before the bending movement. Since the size of the Daimler dataset is relatively small, we intend to use the sliding window method to obtain multiple pedestrian motion sequences to expand the dataset and obtain a substantial amount of data. Finally, the data distribution of each intention category in the unit of sequence after the expansion is shown in Table 3.1.

3.2   INTENTION PREDICTION OF PEDESTRIANS

The proposed framework of the pedestrian intention prediction is shown in Figure 3.2, where the input layer regards the original pedestrian motion image sequence as the system input, and

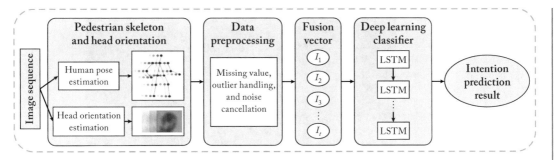

Figure 3.2: Pedestrian intention prediction framework based on machine learning method.

the human pose estimation algorithm is used to extract the characteristic features of the pedestrian skeleton to describe the human motion, meanwhile, a head orientation estimation module is further constructed to improve its prediction accuracy. The skeleton and head orientation features are further preprocessed and combined to reduce the random disturbances and noises of the pedestrian motion, such that, a more delicate posture feature sequence can be obtained. Finally, the fusion feature vector containing the skeleton and head orientation is used as the input of the LSTM deep learning classifier network to predict pedestrian intention. Pedestrian intention prediction network obtains human motion features through skeleton fitting and head orientation estimation systems and then obtains pedestrian intention based on the LSTM network.

3.2.1   EXTRACTION AND PROCESSING OF SKELETON FEATURES

Human pose estimation algorithms are generally divided into two mainstream schemes, namely, top-down and bottom-up human body pose estimation. The top-down human pose estimation scheme [135] is quite a mature pedestrian detection technology and the detection accuracy is relatively high, while the bottom-up human pose estimation scheme [136] is prone to the confusion of different individual skeletons, however, the detection accuracy is relatively low. Thus, we use the mature top-down human pose estimation scheme to obtain pedestrian skeleton features. Also, based on the human body pose estimation algorithm to fit the human skeleton, such that multiple joint points of the human body can be obtained, as shown in Figure 3.3, the skeleton node information at the instant t is denoted by

$$S_t = (a_1, b_1, \ldots, a_k, b_k),\tag{3.1}$$

where (a_u, b_u) denotes the pixel coordinates of the skeleton node with the index k denotes the number of skeleton joint points.

Due to problems such as the low quality of the obtained picture, some parts of the frame skeleton information might be lost during the skeleton fitting process. Therefore, the interpolation method based on the average value of the preceding and following frames is adopted.

Figure 3.3: Schematic diagram of pedestrian skeleton fitting.

Moreover, the difference in the skeleton features of adjacent frames is not obvious when the scale of the pedestrian in an image is small, to alleviate its influence and improve the pedestrian intention prediction performance, the skeleton nodes are normalized within the sequence, namely,

$$a^* = \frac{a - a_{\min}}{a_{\max} - a_{\min}}, \tag{3.2}$$

where $a_{\max}$ and $a_{\min}$ denotes the maximum and minimum values of the horizontal (longitudinal) pixel coordinates of the node in the sequence. In addition, the fitting noise of the skeleton is relatively large due to camera shake and lower image resolution. To alleviate this influence, the normalized skeleton coordinates are subjected to an exponential smoothing filter, namely,

$$\bar{a}_u^t = \beta \sum_{i=0}^{n} (1 - \beta)^i \, a_u^{t-i}, \tag{3.3}$$

where $\bar{a}_u^t$ denotes the skeleton coordinate after filtering at the instant t, n is used to define the size of the filter window, β denotes the damping factor, which is used to control the weight of the historical input information. For instance, the filtering result of the right knee joint coordinates is shown in Figure 3.4.

3.2.2   EXTRACTION AND PROCESSING OF HEAD ORIENTATION

The head orientation of a pedestrian can be defined by the value of its head rotation angle, and the value range is $\omega \epsilon (0, 2\pi)$, thus the head orientation estimation can be regarded as a regression problem. For two-dimensional images, when a pedestrian's head rotates slightly, the difference in image characteristics before and after the rotation is extremely small, but the corresponding

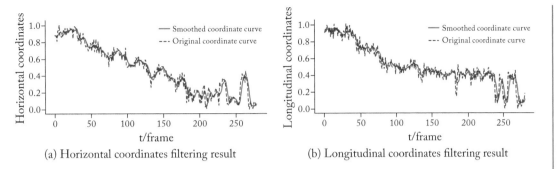

(a) Horizontal coordinates filtering result   (b) Longitudinal coordinates filtering result

Figure 3.4: The effect of filtering the coordinates of the right knee joint.

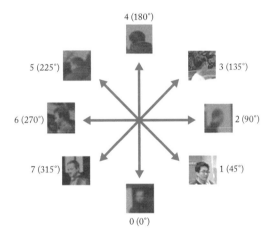

Figure 3.5: Division of the head orientation of pedestrians.

head orientation angles are not consistent. Currently, the regression of the head orientation angle is difficult to obtain. To simplify the concept of pedestrian head orientation to facilitate the task of head orientation estimation, the pedestrian head orientation can be discretized. As shown in Figure 3.5, the head orientation is divided into eight categories and represented by 0–7, where the head orientation angles of two adjacent categories are different by 45°. This chapter uses the head orientation estimation algorithm based on the SVM to obtain the head orientation: first, the histogram feature of the direction gradient of the pedestrian's head is extracted, then the SVM is applied to classify the head orientation result, and finally, the head orientation feature is filtered and smoothed to reduce the disturbance noise [137, 138].

Since the SVM is a classical binary classification model, while the head orientation involved in this research has in total of eight categories, thus, when one direction is selected as the positive category, and the rest of the directions would be regarded as the negative one. That means, eight SVM classifiers are needed to build. Given the pedestrian's head image F_t at the instant t, the head orientation probability can be expressed as

$$P\left(O_t^d\right) = SVM(HOG\left(F_t\right),\ d = 0, 1, \ldots, 7 \tag{3.4}$$

where $P\left(O_t^d\right)$ represents the probability that the *HOG* feature of the image F_t belongs to the head orientation category d, and then the predicted head orientation category O_t is outputted by the *argmax* function:

$$O_t = \text{argmax}\ P\left(O_t^d | F_t\right). \tag{3.5}$$

However, the multi-classifier training can skew the data, that is, there may be samples whose categories cannot be determined. For this type of sample, we should first mark it as an abnormal category. Since the pedestrian motion image is a continuous sequence, the abnormal category can be eliminated when post-processing the data, and then the average value of the head direction category of the preceding and following frames is used as the estimated category of what we mark. In addition, there is no sudden change in the direction of the pedestrian's head in reality. To make the head orientation prediction result in line with the pedestrian's real movement changes, the particle filtering is further used for smoothing:

$$\widehat{O}_t = \sum_{i=1}^{N} \widehat{w}_i^t q_i^t, \tag{3.6}$$

where $\widehat{O}_t$ is the filtering result of the head orientation at the instant t, N denotes the number of particles, q_i^t denotes the direction represented by the i-th particle, and $\widehat{w}_i^t$ denotes the normalized weight, which can be obtained by the following formula:

$$w_i^t = w_i^{t-1} \frac{1}{\sqrt{2\pi\sigma^2}} \exp\left(-\frac{1}{2\sigma^2}\left(q_i^t - O_t\right)\right) \tag{3.7}$$

$$\widehat{w}_i^t = w_i^t / \sum_{K=1}^{N} w_k^t, \tag{3.8}$$

where w_i^t denotes the importance weight of the corresponding particle and σ denotes the variance of the Gaussian distribution probability density function.

3.2.3 FEATURE FUSION METHOD

The pedestrian intention prediction framework can strengthen the representation of pedestrian motion characteristics by combining the skeleton and head orientation features, thereby improving the prediction accuracy of pedestrian intention. Therefore, choosing an appropriate feature fusion method is very important for the LSTM network to extract contextual information. Since the skeleton feature and the head orientation feature have a correspondence relationship, the vector cascade method can be used to fuse the two parts of the feature to obtain a multi-feature sequence I. The fusion feature at time t is:

$$I_t = \left(a_1^t, b_1^t, \ldots, a_u^t, b_u^t, \ldots, a_k^t, b_k^t, O_t\right), \tag{3.9}$$

where (a_u^t, b_u^t) is the coordinate of the skeleton node with index u at the instant t, k denotes the number of skeleton joint points, and O_t denotes the head orientation at the instant t. Since the skeleton sequence is normalized, the range of the skeleton coordinates stays within $[0, 1]$. Meanwhile, the value of the head orientation is $\{0, 1, 2, \ldots, 7\}$, and the numbers of the two types of features are expressed in the same magnitude, so the cascade method can characterize the movement characteristics of pedestrians at that moment to a certain extent.

Although the numbers of the skeleton feature and the head orientation feature are in the same magnitude, the skeleton feature occupies a dominant position after fusion due to a large number of skeleton nodes. To enhance the representation of the head direction feature, the head direction feature is cascaded after each skeleton node coordinate, namely

$$I_t = \left(a_1^t, b_1^t, O_t, \ldots, a_u^t, b_u^t, O_t, \ldots, a_k^t, b_k^t, O_t\right). \tag{3.10}$$

3.2.4   LSTM PEDESTRIAN INTENTION PREDICTION NETWORK BASED ON MULTIPLE FEATURES

To improve the prediction performance of the LSTM network of pedestrian intention, this chapter constructs a pedestrian intention prediction framework that combines pedestrian skeleton with head orientation, as shown in Figure 3.6. The input of this framework is the original pedestrian motion image sequence F

$$F = (F_1, F_2, \ldots, F_{t-1}, F_t), \tag{3.11}$$

where F_t denotes the pedestrian motion image at the instant t. The top-down human pose estimation algorithm is used to obtain the pedestrian skeleton features, and the missing value supplement, normalization, and exponential smoothing filtering are performed on them to obtain a delicate skeleton sequence:

$$S = (S_1, S_2, \ldots, S_{t-1}, S_t). \tag{3.12}$$

The head direction estimation algorithm is used to extract the head direction features of pedestrians. Also, the abnormal value removal and particle filter processing are performed on them to obtain the head direction sequence that conforms to the law of pedestrian movement.

$$O = (O_1, O_2, \ldots, O_{t-1}, O_t). \tag{3.13}$$

The preprocessed skeleton and head orientation are fused to obtain a fusion vector sequence I:

$$I = (I_1, I_2, \ldots, I_{t-1}, I_t), \tag{3.14}$$

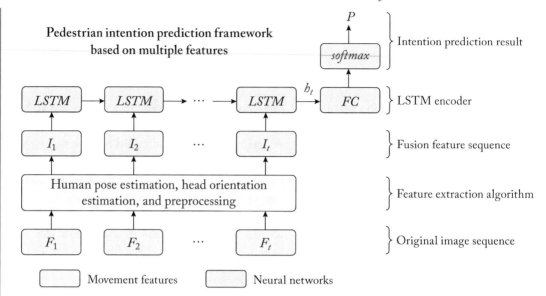

Figure 3.6: Pedestrian intention prediction framework.

where $I_t = (S_t, O_t)$, S_t and O_t, respectively, represent the skeleton node vector and head orientation at the instant t. In the LSTM encoder, the fusion vector sequence I first passes through the LSTM to obtain the intention coding vector h_t that combines the head orientation and skeleton information. Then, the fully connected layer is used to map h_t to the sample label space. Finally, the probability distribution of intention is obtained by the softmax function:

$$P = \text{softmax}\left(W_s h_t + b_s\right), \tag{3.15}$$

where W_s and b_s denote the weights and biases of the fully connected layer, respectively, and the probability distribution P is

$$P = (P_1, P_2, \ldots, P_n), \tag{3.16}$$

where n denotes the total number of pedestrian intention categories and P_n represents the probability that the fusion vector sequence I belongs to the intention category n. The predicted intention category can be calculated by the *argmax* function. Moreover, the cross-entropy loss function is used to optimize the weight and bias of the intention prediction network, namely:

$$H(Q, P) = \sum_i Q(i) \log\left(\frac{1}{P(i)}\right), \tag{3.17}$$

where $P(i)$ and $Q(i)$ denote the predicted probability distribution and the true probability distribution, respectively.

3.3   EXPERIMENT AND ANALYSIS

For intention prediction methods, the accuracy, precision, recall, and F1 scores can be used as performance evaluation indicators.

Accuracy is defined as

$$\text{Accuracy} = \frac{TP + TN}{TP + TN + FP + FN},$$

(3.18)

where TP represents the number of samples that are positive and predicted to be positive, TN represents the number of samples that are negative and predicted to be negative, FP represents the number of samples that are negative and predicted to be positive, and FN represents the number of samples that are positive and predicted to be negative. It can be seen that accuracy is the proportion of the number of samples with correct predictions to the total number of samples, which can be used to measure the overall performance of the model.

Precision is defined as

$$\text{Precision} = \frac{TP}{TP + FP}.$$

(3.19)

It can be seen that precision represents the fraction of positive samples that are predicted to be positive among all samples that are predicted to be positive, which can evaluate the model's ability to discriminate negative samples.

The recall is defined as

$$\text{Recall} = \frac{TP}{TP + FN}.$$

(3.20)

The recall is the fraction of positive samples that are predicted to be positive among all positive samples, which can evaluate the model's ability to recognize positive samples.

F1 score is defined as

$$\text{F1\_Score} = \frac{2 * \text{Precision} * \text{Recall}}{\text{Precision} + \text{recall}}.$$

(3.21)

It can be noticed from its definition that the F1 score is the harmonic average of precision and recall. Therefore, when the actual task is sensitive to both precision and recall, the F1 score can measure the overall performance of the model.

When introducing the head orientation features, we construct two intention prediction models to verify the impact on the performance of intention prediction: (i) a single feature model that only considers pedestrian skeleton information (Sk); and (ii) a multi-featured model that considers skeleton and head orientation (Sk + He). To explore the influence of historical information length on the performance of intention prediction, the above two models were trained with features of different historical lengths, and the accuracy of intention prediction obtained is shown in Table 3.2. It can be seen that the accuracy of intention prediction is significantly improved as the sequence length increases but is not larger than 32. Thus, considering the network calculation cost and prediction accuracy, this study uses a historical sequence of length 32 to train and test the model.

Table 3.2: The accuracy of intention prediction under different sequence lengths

Sequence Length	Single Feature Model	Multi-Feature Model
8	0.583	0.786
16	0.702	0.893
24	0.877	0.933
32	0.925	0.960
40	0.921	0.960
48	0.913	0.940

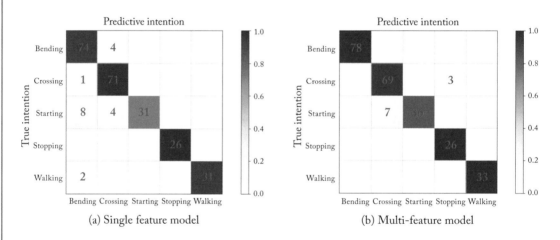

(a) Single feature model   (b) Multi-feature model

Figure 3.7: Intention prediction confusion matrix.

The confusion matrices of the single-feature model and the multi-feature model are shown in Figure 3.7. The number of prediction error samples of the multi-feature model is significantly less than that of the single-feature model. Precision, recall, F1 score, and accuracy can be further calculated from the confusion matrix, and the results are shown in Table 3.3, which shows that the intention prediction accuracy of the single- and multi-feature models is 0.925 and 0.960, respectively. Thus, the introduction of head orientation can enhance the expression of pedestrian motion features and improve the overall performance of intention prediction.

Moreover, based on the accuracy, recall rate, and F1 score, we can conclude that the multi-feature model can significantly improve the prediction performance of bending, and the reason could be that a pedestrian is usually facing the camera or having their back to it before bending. While in all other intentions, the pedestrian is the side-facing camera for a longer time, as a result, the introduction of head features helps the network to better distinguish between bending and other intentions. In addition, crossing and stopping are slightly lower in F1 scores and other

Table 3.3: Pedestrian intention prediction performance

		Precision		Recall		F1_Score		Accuracy	
Features		Sk	Sk+He	Sk	Sk+He	Sk	Sk+He	Sk	Sk+He
Intention categories	Bending	0.871	1.000	0.949	1.000	0.908	1.000	0.925	0.960
	Crossing	0.899	0.908	0.986	0.958	0.940	0.932		
	Starting	1.000	1.000	0.721	0.837	0.838	0.911		
	Stopping	1.000	0.897	1.000	1.000	1.000	0.946		
	Walking	1.000	1.000	0.939	1.000	0.969	1.000		

Remarks: Sk (skeleton feature), He (head feature)

indicators. The reason is that the head orientation of pedestrians in these two types of intentions becomes one source of interference, people may observe their left and right without changing their intentions when crossing and stopping.

Pedestrian intentions in actual scenes can change from walking to bending, or from crossing to stopping. Foreseeing the changes in pedestrian intentions will be helpful for intelligent vehicle decision-making and planning, thereby improving driving safety. We will verify the effectiveness of the proposed framework by comparing it with the single- and multi-feature models when dealing with intention changes. These results are shown in Figures 3.8 and 3.9, respectively, in which the horizontal axis is time to event (TTE) that represents the period from the peak point of the pedestrian's intention change. For example, TTE = 0 is the exact finishing point of the pedestrian's intention change; TTE > 0 means the period before the pedestrian's intention change, while TTE < 0 represents the period after the intention change. As shown in Figure 3.8, with 0.5 as the classification threshold, the single-feature and multi-feature models can predict the pedestrian intention change from walking to bending by 2 and 9 frames in advance, respectively. The result indicates that the introduction of the head orientation can increase pedestrian intention predicted time from 0.13–0.56 s (the acquisition frequency is 16 frames per second). Moreover, as shown in Figure 3.9, the prediction time of the single-feature and multi-feature models are 0.06 s (1 frame) and 0.19 s (3 frames), respectively, during the transition from traversing to stopping. The multi-feature model has a limited lifting effect in this scenario, which might be related to insignificant changes in the head orientation of pedestrians during the transition.

3.4 TRAJECTORY PREDICTION OF PEDESTRIANS

Before moving to the trajectory prediction of the pedestrian, first, we propose a preprocessing method for vehicle displacement compensation and coordinate translation transformation, which is based on the features of pedestrian trajectory noise in vehicle frontal view scene. Because

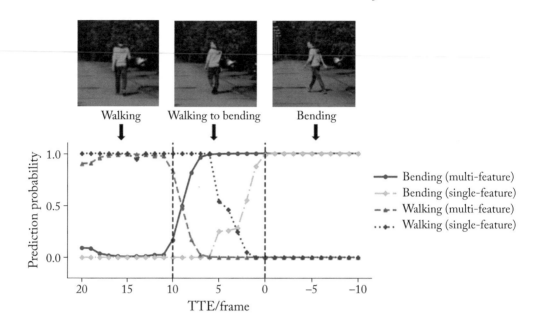

Figure 3.8: Intention probability changes from walking to bending.

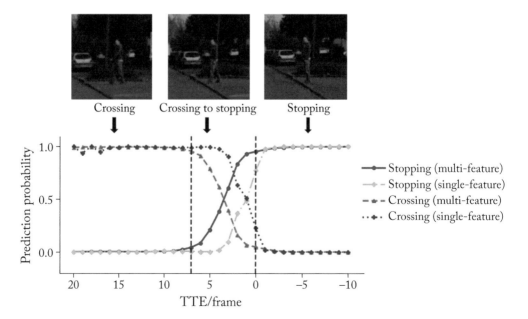

Figure 3.9: Intention probability changes from crossing to stopping.

Early position   Mid-term position   Late position

Figure 3.10: Schematic diagram of pedestrian position coordinate collection.

of the uncertainty of pedestrian movement and the difficulty of the kinematics model to accurately describe the dynamic change of pedestrians, one LSTM trajectory prediction network based on the enhanced attention mechanism is constructed. Considering that the pedestrian motion trajectory could be regarded as a concrete projection of his/her intention, in reality, a hierarchical pedestrian trajectory prediction framework incorporating pedestrian intentions is further proposed to improve the trajectory prediction performance.

3.4.1   CHARACTERISTICS AND PREPROCESSING METHODS OF PEDESTRIAN TRAJECTORY DATA

The position coordinates of the pedestrian in the frontal view of the vehicle are collected by the radar, and the position coordinates of the pedestrian will be related to the movement of the vehicle, which brings difficulties to the trajectory prediction of the pedestrian. As shown in Figure 3.10, a complete bending sequence includes three stages, namely, the early, middle, and late phases. If the early phase is treated as the reference, the interference for the pedestrian trajectory due to the vehicle movement would become noticeable with time, also, the varying vehicle speed could worsen the situation.

As shown in Figure 3.11, to restore the true position of the pedestrian relative to the ground, vehicle displacement compensation processing is executed to reduce the interference from the relative movement of pedestrians and vehicles, namely

$$\begin{cases} x_j^p = \check{x}_j^p + x_j^v \\ y_j^p = \check{y}_j^p + y_j^v, \end{cases} \tag{3.22}$$

where x_j^p and y_j^p are the horizontal and longitudinal coordinates of the pedestrian relative to the initial reference position of the vehicle at time j, $\check{x}_j^p$, and $\check{y}_j^p$ are the horizontal and longitudinal coordinates of the pedestrian relative to the vehicle at time j, x_j^v, and y_j^v is the horizontal and longitudinal distance the vehicle travels from the initial reference time to time j.

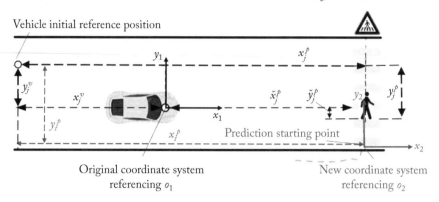

Figure 3.11: Schematic diagram of pedestrian trajectory data processing.

In addition, when the relative distance between pedestrians and vehicles differs much, the corresponding trajectory coordinates will also differ greatly, which will bring difficulties to the network training, such that, further reducing the network convergence speed. To improve the performance of trajectory prediction, we regard the starting point of the trajectory prediction as the reference origin to perform coordinate translation transformation, namely,

$$\begin{cases} x_j = x_j^P - x_t^P \\ y_j = y_j^P - y_t^P, \end{cases} \tag{3.23}$$

where the lower-index t denotes the time corresponding to the starting point of the trajectory prediction, x_t^P and y_t^P are the horizontal and longitudinal coordinates of the pedestrian relative to the initial reference position of the vehicle at that time, and x_j and y_j are the horizontal and longitudinal coordinates after translational transformation at time t.

As shown above, the historical trajectory can be processed by vehicle displacement compensation and coordinate translation transformation to alleviate those interference noises, thereby improving the prediction performance of the trajectory prediction model for the future trajectory of pedestrians.

3.4.2 PEDESTRIAN TRAJECTORY PREDICTION BASED ON THE KINEMATICS MODEL

The state of a pedestrian can be described by kinematic features such as speed and position. Kalman filter (KF) is a linear optimal filter that can estimate the position of the pedestrian by constructing a state transition equation of the states and noises, which can be used for realizing the pedestrian trajectory prediction based on the kinematics model [139].

(1) *Kinematics model*:   The kinematics models commonly used by pedestrians in the frontal view scene of the vehicle include the constant velocity (CV) model and the constant position (CP) model, in which the CV model describes pedestrian motion as walking in a straight line with a uniform speed, and its acceleration is zero, and its kinematics equation is

$$X_t = \begin{bmatrix} x + v_x \cdot \Delta t \\ y + v_y \cdot \Delta t \\ v_x \\ v_y \end{bmatrix}_{t-1} + \begin{bmatrix} \frac{\Delta t^2}{2} & 0 \\ \frac{\Delta t^2}{2} & 0 \\ \Delta t & 0 \\ 0 & \Delta t \end{bmatrix} \begin{bmatrix} \mathcal{T}_{a_x} \\ \mathcal{T}_{a_y} \end{bmatrix}, \tag{3.24}$$

where x and y denote the horizontal and longitudinal positions of the pedestrian, v_x and v_y denote the horizontal and longitudinal speeds, Δt denotes the time interval, and $\mathcal{T}_{a_x}$ and $\mathcal{T}_{a_y}$ denote the acceleration interference terms. The CP model assumes that the pedestrian position is constant and its speed is zero, so the kinematics equation can be expressed as:

$$X_t = \begin{bmatrix} x \\ y \\ 0 \\ 0 \end{bmatrix}_{t-1} + \begin{bmatrix} \Delta t & 0 \\ \Delta t & 0 \\ 1 & 0 \\ 0 & 1 \end{bmatrix} \begin{bmatrix} \mathcal{T}_{v_x} \\ \mathcal{T}_{v_y} \end{bmatrix}, \tag{3.25}$$

where $\mathcal{T}_{v_x}$ and $\mathcal{T}_{v_y}$ denote the horizontal and longitudinal velocity interference terms.

(2) *Kalman filter based on the single motion model*:   KF obtains the estimated value of the target variable (pedestrian position) through the state equation and then obtains the measured value through the observation equation to correct the estimated value, thereby obtaining the optimal estimate of the target variable. For a given kinematic model, the state equation is represented by

$$X_t = AX_t + W_t, \tag{3.26}$$

where X_t denotes the estimated value of the state variable at time t, A denotes the state transition matrix, and W_t denotes the Gaussian white noise, which follows the following distribution

$$p(W) \sim N(0, Q), \tag{3.27}$$

where Q denotes the covariance matrix. The observation equation is

$$Z_t = HX_t + V_t, \tag{3.28}$$

where Z_t denotes the measured value of the system at time t, H denotes the observation matrix of the kinematical model, and V_t denotes the Gaussian white noise, which follows the following distribution

$$p(V) \sim N(0, R), \tag{3.29}$$

where R denotes the covariance matrix.

Then the Kalman filtering process by the state equation and the observation equation is as follows.

First, obtain the estimated value of the state variable $X_{t|t-1}$ and the prior probability $P_{t|t-1}$:

$$X_{t|t-1} = A X_{t-1|t-1} + W_t \tag{3.30}$$

$$P_{t|t-1} = A P_{t-1|t-1} + Q, \tag{3.31}$$

where $X_{t-1|t-1}$ denotes the optimal estimation of the state variable at time $t-1$, $P_{t-1|t-1}$ denotes the posterior probability at time $t-1$.

Then calculate the Kalman gain K_t by

$$K_t = P_{t|t-1} H^T \left(H P_{t|t-1} H^T + R \right)^{-1}. \tag{3.32}$$

Finally, update the estimated value of the state variable and the predicted covariance matrix:

$$X_{t|t} = X_{t|t-1} + K_t \left(Z_t - H X_{t|t-1} \right) \tag{3.33}$$

$$P_{t|t} = (I - K_t H) P_{t|t-1}, \tag{3.34}$$

where $X_{t|t}$ denotes the optimal estimation of the state variable at time t, and $P_{t|t}$ denotes the posterior probability at time t, and $X_{t|t}$ and $P_{t|t}$ would be passed to the next time-step to repeat the Kalman filtering process.

(3) *Kalman filter based on interacting multiple models*:   Since the walking state of pedestrians may change, such as from crossing intention to stopping intention, the interacting multiple models (IMM) [140] uses multiple model filters to estimate the state of the target, and each filter has a weight coefficient. The filter is updated through the probability transition matrix, and the optimal estimated value is obtained by the weighted calculation of multiple filter state estimates. IMM is expected to improve the prediction results of the KF.

Suppose there are k kinematic models in the IMM algorithm, and the system state equation and system observation equation are:

$$X_t = A_i X_t + W_t^i, \; i = 1, 2, 3, \ldots, k \tag{3.35}$$

$$Z_t = H_i X_t + V_t^i, \tag{3.36}$$

where A_i and H_i denotes the state transition matrix and the observation matrix of the i-th model filter, W_{it} and V_{it} denotes Gaussian white noise, and the corresponding covariance matrixes are Q_i and R_i, respectively. The model probability transition matrix of IMM is

$$p = \begin{bmatrix} p_{11} & \cdots & p_{1k} \\ \vdots & \ddots & \vdots \\ p_{k1} & \cdots & p_{kk} \end{bmatrix}, \tag{3.37}$$

where p_{ij} represents the transition probability between the model i and j.

According to the above definition, the state estimation process of the IMM can be carried out, which includes a total of four steps, namely, model interaction, filter filtering, probability update, and data fusion.

1. **Model interaction**: the initial conditions $X^{0i}_{t-1|t-1}$ and $P^{0i}_{t-1|t-1}$ after model interaction can be obtained by the following formula:

$$X^{0i}_{t-1|t-1} = \sum_{j=1}^{k} X^i_{t-1|t-1} u^{ij}_{t-1|t-1}, \quad i = 1, 2, 3, \ldots, k \tag{3.38}$$

$$P^{0i}_{t-1|t-1} = \sum_{j=1}^{k} u^{ij}_{t-1|t-1} \tag{3.39}$$
$$\left[P^i_{t-1|t-1} + \left(X^i_{t-1|t-1} - X^{0i}_{t-1|t-1} \right) \left(X^i_{t-1|t-1} - X^{0i}_{t-1|t-1} \right)^T \right],$$

where $u^{ij}_{t-1|t-1}$ is the interaction weight between the model i and model j, which can be calculated by the following formula:

$$u^{ij}_{t-1|t-1} = p_{ij} u^j_{t-1} / \bar{c}_i \tag{3.40}$$

$$\bar{c}_i = \sum_{j=1}^{k} p_{ij} u^j_{t-1}. \tag{3.41}$$

2. **Filtering process**: Calculate the state equation and predict covariance matrix from the initial conditions of the model interaction:

$$X^i_{t|t-1} = A_i X^{0i}_{t-1|t-1} + W^i_t \tag{3.42}$$

$$P^i_{t|t-1} = A_i P_{t-1|t-1} + Q_i, \tag{3.43}$$

where $X^i_{t|t-1}$ denotes the optimal estimation of the state variable of the i-th model at time $t-1$, and $P^i_{t|t-1}$ denotes the prior probability corresponding to the i-th model. Then calculate the Kalman gain K^i_t by

$$K^i_t = P^i_{t|t-1} H^T_i \left(H_i P^i_{t|t-1} H^T_i + R_i \right)^{-1}. \tag{3.44}$$

The estimated value of the state variable and the predicted covariance matrix can be updated by

$$X^i_{t|t} = X^i_{t|t-1} + K^i_t \left(Z^i_t - H_i X^i_{t|t-1} \right) \tag{3.45}$$

$$P^i_{t|t} = \left(I - K^i_t H_i \right) P^i_{t|t-1},$$ (3.46)

where $X^i_{t|t}$ denotes the optimal estimation of the state variable of the i-th model at time t and $P^i_{t|t}$ denotes the posterior probability of the i-th model at time t.

3. **Probability update**: The IMM algorithm uses the maximum likelihood function to characterize the similarity between the current model and the target motion state, thereby updating the model probability by

$$\Lambda^i_t = \frac{1}{\sqrt{2\pi \left| S^i_t \right|}} \exp \left(-\frac{1}{2} d^{i\,T}_t S^{i\,-1}_t d^i_t \right),$$ (3.47)

where Λ^i_t denotes the maximum likelihood function of the target motion state at time t and the model i, S^i_t and d^i_t can be obtained by the following formula:

$$S^i_t = H P^i_{t|t-1} H^T + R_i$$ (3.48)

$$d^i_t = Z^i_t - H_i X^i_{t|t-1}.$$ (3.49)

The updated probability of the model can be obtained from the likelihood function:

$$u^i_t = \frac{1}{c} \Lambda^i_t \sum_{j=1}^{k} p_{ij} u^j_{t-1},$$ (3.50)

where c is the normalization constant, which can be obtained by the following formula:

$$c = \sum_{j=1}^{k} \Lambda^i_t \bar{c}_i.$$ (3.51)

4. **Data fusion**: The optimal estimated value of the overall system is obtained from the estimated value of each model filter and its corresponding weight:

$$X_{t|t} = \sum_{i=1}^{k} X^i_{t|t} u^i_t.$$ (3.52)

And the overall covariance is

$$P_{t|t} = \sum_{i=1}^{k} u^i_t \left[P^i_{t|t} + \left(X^i_{t|t} - X_{t|t} \right) \left(X^i_{t|t} - X_{t|t} \right)^T \right],$$ (3.53)

where the overall optimal estimated value $X_{t|t}$ and the posterior probability $P_{t|t}$ can be used as the next interactive input, thereby recursively completing the entire prediction process.

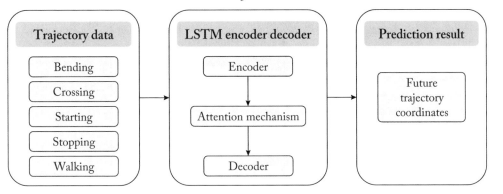

Figure 3.12: Pedestrian trajectory prediction framework based on enhanced attention mechanism.

3.4.3   LSTM PEDESTRIAN TRAJECTORY PREDICTION NETWORK BASED ON THE ENHANCED ATTENTION MECHANISM

This section uses the LSTM network encoder-decoder to construct a pedestrian trajectory prediction framework based on the enhanced attention mechanism, as shown in Figure 3.12. The framework learns the dynamic characteristics of pedestrian trajectories through the encoder and introduces an attention mechanism to enhance the effective usages of the pedestrian historical state information. Finally, the decoder adopts the enhanced features from the attention mechanism to generate future trajectory coordinates.

(1) *Encoder-decoder based on LSTM network*:   The LSTM pedestrian trajectory prediction network built with the encoding-decoding framework is shown in Figure 3.13. The network consists of the input layer, LSTM encoder, LSTM decoder, and output layer.

The input of the network is the historical trajectory coordinate sequence X:

$$X = (X_1, X_2, \ldots, X_{t-1}, X_t), \tag{3.54}$$

where $X_t = (x_t, y_t)$, x_t and y_t denote the pre-processed pedestrian position coordinates at time t. Then the sequence X gets the code vector h_t through the LSTM encoder, such that,

$$\widetilde{X}_t = W_{FC} * X_t + b_{FC} \tag{3.55}$$

$$h_t = LSTM\left(h_{t-1}, c_{t-1}, \widetilde{X}_t, W_e, b_e\right), \tag{3.56}$$

where W_{FC} and b_{FC} denote the weight and offset corresponding to the fully connected layer of the encoder, respectively, which are used to map the coordinate X_t into a high-dimensional space to enhance the nonlinear fitting ability of the encoder, W_e and b_e are the weights and offsets of the LSTM encoder, respectively, which are used to extract the time series feature information contained in the high-dimensional vector $\widetilde{X}_t$, h_t is the hidden state at the last moment of the

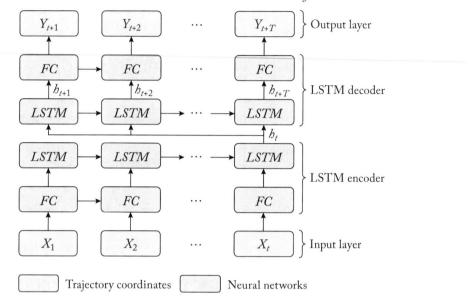

Figure 3.13: LSTM pedestrian trajectory prediction network.

LSTM unit, which contains the context information of the pedestrian historical coordinates, and it is used as the input of the LSTM decoder, namely,

$$h_\tau = LSTM(h_{\tau-1}, c_{\tau-1}, h_t, W_d, b_d), \quad (3.57)$$

where W_d and b_d denote the corresponding weights and offsets of the LSTM decoder, respectively, h_τ denotes the output of the LSTM decoder at time τ. The output of the LSTM unit at each time are concatenated to obtain the decoded vector sequence H_d, namely,

$$H_d = (h_{t+1}, h_{t+2}, \ldots, h_{t+T-1}, h_{t+T}). \quad (3.58)$$

The fully connected layer is used to map the decoded vector sequence H_d into the sample label space to obtain the pedestrian future trajectory coordinate sequence Y, such that

$$Y = (Y_{t+1}, Y_{t+2}, \ldots, Y_{t+T-1}, Y_{t+T}), \quad (3.59)$$

where $Y_{t+T} = (x_{t+T}, y_{t+T})$ represents the trajectory coordinates of pedestrians at time T in the future.

(2) *Attention mechanism*:   As shown in Equation (3.57), the decoder usually takes the last output of h_t as the encoding vector. However, the output of the LSTM unit at other moments also contains the characteristics of the early motion state of the pedestrian. To obtain more information about pedestrian movement trends, we introduce an attention mechanism in the

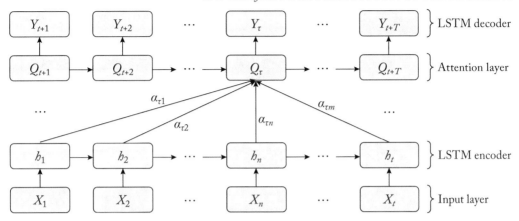

Figure 3.14: Schematic diagram of attention mechanism.

encoder-decoder to address the output of the LSTM at each moment, which can enhance the performance of the LSTM network and improve the accuracy of trajectory prediction [141]. As shown in Figure 3.14, the hidden state of input X can be obtained at each instant through the LSTM encoder, and the output sequence H_e can be represented by

$$H_e = (h_1, h_2, \ldots, h_n, \ldots, h_t).\tag{3.60}$$

The attention layer dynamically assigns weights to the output of the encoder at each moment, and aligns the code vector sequence Q with the output of the decoder:

$$Q = (Q_{t+1}, Q_{t+2}, \ldots, Q_\tau, \ldots, Q_{t+T}),\tag{3.61}$$

where Q_τ denotes the weighted sum of LSTM output at each time, namely,

$$Q_\tau = \sum_{n=1}^{t} \alpha_{\tau n} h_n,\tag{3.62}$$

where $\alpha_{\tau n}$ denotes the weight of the code vector at time τ relative to the encoder output at the instant n, which can be calculated by

$$\alpha_{\tau n} = \exp(e_{\tau n}) / \sum_{k=1}^{t} e_{\tau k}\tag{3.63}$$

$$e_{\tau n} = g(h_n),\tag{3.64}$$

where $e_{\tau n}$ denotes the importance of h_n on Q_τ, and $g(\ )$ is the fully connected network.

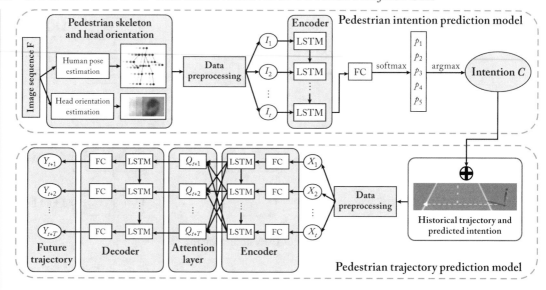

Figure 3.15: Hierarchical pedestrian trajectory prediction framework incorporating intentions.

3.5 EVALUATION OF THE HIERARCHICAL PEDESTRIAN TRAJECTORY PREDICTION FRAMEWORK INCORPORATING PEDESTRIAN INTENTIONS

Pedestrian movements are usually uncertain, and their trajectories could be regarded as the projection of their intention in reality. To consider the influence of intention on pedestrian movement and improve the prediction performance of pedestrian trajectory, we propose a hierarchical pedestrian trajectory prediction framework that integrates pedestrian intention. The intention prediction model is used to enhance the pedestrian motion characteristics, then, along with the historical trajectory, they will be regarded as the input feature of the trajectory prediction network to generate the future trajectory.

The hierarchical pedestrian trajectory prediction framework incorporating the pedestrian intention is shown in Figure 3.15, which is composed of a pedestrian intention prediction module and a pedestrian trajectory prediction module. The input of the entire frame is the original image sequence $F = (F_1, F_2, \ldots, F_{t-1}, F_t)$ and the pedestrian historical trajectory coordinate sequence $\widehat{X} = (\widehat{X}_1, \widehat{X}_2, \ldots, \widehat{X}_{t-1}, \widehat{X}_t)$.

The intention prediction module has been well introduced in Section 3.3, while the trajectory prediction module is based on the LSTM with the attention mechanism as introduced in Section 3.4.3. Due to the large difference in the vector dimension between the intention feature and the historical trajectory coordinate sequence, a cascade replication method is used to fuse the intention and the historical trajectory coordinates to enable the trajectory prediction network to learn the two types of features in a balanced manner. The preprocessed input sequence X of the trajectory predictive encoder can be obtained by

$$X = (X_1, X_2, \ldots, X_{t-1}, X_t), \qquad (3.65)$$

where $X_t = (x_t, y_t, C)$, x_t, and y_t represent the pedestrian trajectory coordinates after vehicle displacement compensation and coordinate translation transformation at time t, respectively, and C denotes the intention prediction category corresponding to the trajectory sequence. The trajectory coding vector sequence Q is obtained through the encoder and the attention layer of the trajectory prediction module, and then it will be used to generate the future trajectory coordinate sequence Y through the decoder.

3.5.1   EXPERIMENT AND ANALYSIS

A previous study has shown that 90% of rear-end collisions and 60% of frontal collisions can be avoided if the driver can realize the potential traffic danger 1 s in advance and take corresponding measures [142]. Moreover, considering the variability of pedestrian movement in the long prediction window, we only evaluate the prediction results of pedestrian trajectory within 1 s. To measure the overall performance of the model, Root Mean Square Error (RMSE) is selected as the evaluation index, because the RMSE could amplify the error caused by the abnormal prediction value. The calculation of the RMSE is

$$RMSE = \sqrt{\frac{1}{MT} \sum_{m=1}^{M} \sum_{i=1}^{T} \left(x_m^i - \hat{x}_m^i\right)^2 + \left(y_m^i - \hat{y}_m^i\right)^2}, \qquad (3.66)$$

where M denotes the number of training samples, T denotes the total time length of the trajectory prediction, $(x_m^i, y_m^i$ and $(\hat{x}_m^i, \hat{y}_m^i)$ are the real trajectory coordinates and predicted trajectory coordinates of the m-th sample at time i. To accomplish the comparative study, we will consider the following pedestrian trajectory prediction models:

1. **CV-KF** [63], in which the CV model and Kalman filter are used to predict pedestrian trajectory.

2. **IMM-KF** [63], which is based on the CV-KF, a CP model is introduced to realize pedestrian trajectory prediction based on the IMM.

3. **LSTM** [67], in which only the historical trajectory is used as the input, and the predicted pedestrian position coordinates are outputted through the encoding and decoding process of the LSTM network.

4. **Intention-based LSTM** (I-LSTM), which is based on the LSTM prediction network with pedestrian intentions to realize the trajectory prediction.

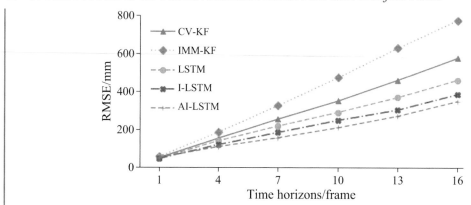

Figure 3.16: Comparison of the root mean square error of each trajectory prediction model.

Table 3.4: Comparison of RMSE (mm) of different models

Method	Time Horizons/Frames					
	1	4	7	10	13	16
CV-KF	47	151	252	348	458	574
IMM-KF	53	183	323	471	626	774
LSTM	43	138	216	289	367	459
I-LSTM	42	117	182	247	301	384
AI-LSTM	48	107	154	208	269	347

5. **Attention and Intention-based LSTM** (AI-LSTM), which is based on the I-LSTM, the attention mechanism is introduced to enhance the performance of the LSTM network and realize pedestrian trajectory prediction. This is a complete hierarchical pedestrian trajectory prediction framework incorporating pedestrian intentions that are proposed in this chapter.

The results of each method on the test set are shown in Figure 3.16 and Table 3.4. Figure 3.16 shows that, with the prediction time increases, the RMSE of each model has an upward trend, the performance of the LSTM-based method is better than the method based on the kinematics model, and the prediction performance based on the LSTM improves more significantly than the kinematic model. Therefore, the data-driven model has better performance than the kinematics model when dealing with trajectory prediction problems. That can be also seen from the results shown in Table 3.4, where the RMSE of the I-LSTM is 384 mm when the prediction time is 16 frames. Compared with the LSTM, the I-LSTM can reduce the error by 16.3%, indicating that the hierarchical trajectory prediction framework fused with intention features can significantly improve the trajectory prediction performance, while the AI-LSTM has an RMSE of 347 mm, which is 9.6% lower than that of the I-LSTM, indicating that the enhanced attention mechanism can further improve the performance of trajectory prediction.

There is a change of pedestrian intention in the actual scene. To evaluate the trajectory prediction performance of the proposed model in the scene with intention changes, the following will compare and analyze the performance of LSTM, I-LSTM, and the AI-LSTM under the scenes of walking to bending and crossing to stopping. Since Mean Absolute Error (MAE) can more intuitively measure the deviation of the predicted value from the true value, the following experiment uses MAE as the evaluation index. The MAE is calculated by

$$MAE = \frac{1}{MT} \sum_{m=1}^{M} \sum_{i=1}^{T} \sqrt{\left(x_m^i - \hat{x}_m^i\right)^2 + \left(y_m^i - \hat{y}_m^i\right)^2}, \tag{3.67}$$

where M denotes the number of training samples, T denotes the total time length of the trajectory prediction, and (x_m^i, y_m^i) and $(\hat{x}_m^i, \hat{y}_m^i)$ are the real and predicted trajectory coordinates of the m-th sample at time i.

Figures 3.17 and 3.18 show the error curves of three comparative models (LSTM, I-LSTM, AI-LSTM) at different TTEs, respectively. Table 3.5 shows the MAE at the last moment of each curve in Figures 3.17 and 3.18. It can be observed that the MAE of I-LSTM is smaller than that of the LSTM, indicating that the fusion of intention can improve the performance of trajectory prediction. The MAE of the AI-LSTM is slightly lower than that of the I-LSTM, indicating that the attention mechanism can reduce the trajectory prediction error but the improvement could be limited. In addition, since the attention mechanism can extract more pedestrian movement trend information, the AI-LSTM prediction error curve is smoother than that of the I-LSTM. To further illustrate the influence of fusing the intention information on the performance of trajectory prediction, the following comparative analysis would only focus on the I-LSTM and the LSTM.

Table 3.5 shows that the MAE increases first and then decreases with the growth of the TTE, and the MAE near the peak point of intention change (TTE = 0) is large, which is caused by the change of pedestrian's intention, increasing the uncertainty of the trajectory, thus, it could be difficult to predict the trajectory precisely. As for the trajectory prediction from walking to bending, which is shown in Figure 3.17a and b, as well as in Table 3.5, we can observe that the I-LSTM could predict the pedestrian's bending intention 9 frames in advance (TTE = 9), while the I-LSTM can predict the future trajectory in combination with the bending intention when TTE = 8. Moreover, its maximum error is only 228 mm, which is 50.0% lower than that of the LSTM. Therefore, the hierarchical trajectory prediction framework fused with intention features could reduce the uncertainty of pedestrian movement and improve the predictive performance of pedestrian trajectories in this scene.

Regarding the trajectory prediction from crossing to stop, which is shown in Figure 3.18a, b, and Table 3.5, we can see that the prediction error of I-LSTM is larger than that of the LSTM

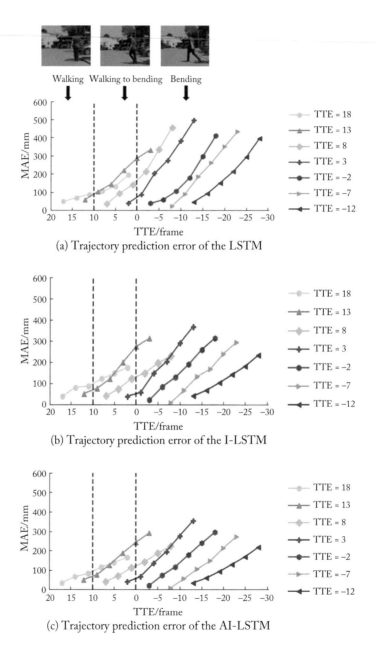

Figure 3.17: The prediction errors at different prediction moments from walking to bending.

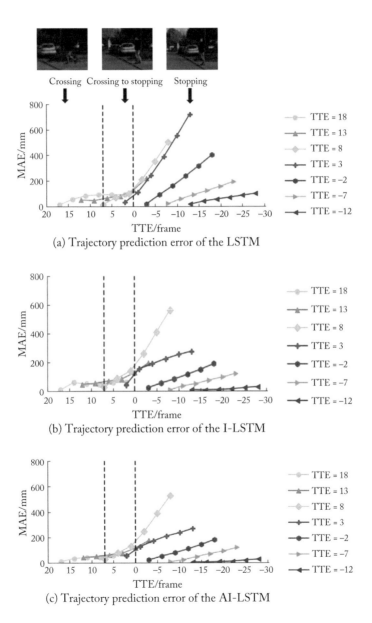

Figure 3.18: The prediction errors at different prediction moments from crossing to stopping.

Table 3.5: Maximum MAE predicted from each TTE (mm)

TTE/Frame	Walking to Bending (mm)			Crossing to Stopping (mm)		
	LSTM	I-LSTM	AI-LSTM	LSTM	I-LSTM	AI-LSTM
18	194	175	164	90	60	57
13	333	312	290	218	193	174
8	456	228	223	504	563	529
3	496	368	353	718	276	266
−2	411	313	294	399	189	182
−7	433	294	273	192	121	118
−12	394	232	217	98	28	26

when TTE = 8, which can be explained from Figure 3.9: the I-LSTM's prediction ability for the stopping during the intention transition is 3 frames (TTE = 3), that means the intention prediction result of TTE = 8 is still a traverse, the fusion of this intention would increase the trajectory prediction error. When TTE = 3, the I-LSTM can predict the transition from crossing to stopping. Currently, its maximum MAE is only 276 mm, which is 61.6% lower than the LSTM. It shows that the fusion of intention information in the trajectory prediction network can effectively reduce the impact of intention change, and significantly reduce the trajectory prediction error in an intention change scene.

3.5.2   CASE STUDY

In this section, we will use the proposed framework to predict pedestrian intentions for a real accident scene to illustrate its effectiveness to prevent potential collision accidents. The actual scene of the accident is shown in Figure 3.19, while the top view of the collision accident scene is depicted in Figure 3.20, in which a car collided with a pedestrian on New Songjiang Road on May 17, 2018, resulting in the death of a pedestrian. Before the accident, the car move normally on the road at a speed of 45 km/h, and the pedestrians crossed the road at a speed of 8 km/h. However, the pedestrians did not pay attention to the traffic situation, and the driver failed to notice them timely. Thus, a collision between the car and the pedestrian happened. After the collision, the driver began to take the braking maneuver, and the braking distance was 18.5 m. Table 3.6 supplied parameter explanations and values as appeared in Figure 3.20. To reconstruct the traffic scene before the accident, we make the following assumptions:

1. The car travels at a constant speed of 45 km/h from east to west.

2. The pedestrian is still at the beginning and then crossed the road at a constant speed of 8 km/h.

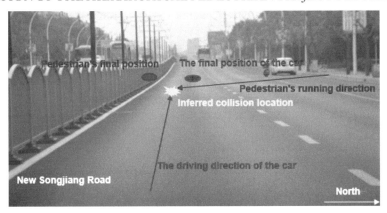

Figure 3.19: The scene of the collision accident.

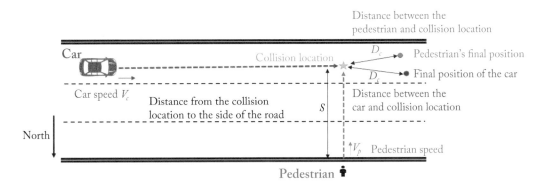

Figure 3.20: The top view scene of the collision accident scene.

3. Due to distraction, the vehicle driver did not take braking before the collision.

4. Pedestrians did not notice the car before the collision.

Based on the above assumptions and the parameters shown in Table 3.6, we used the Prescan® to rebuild the scenario of the pedestrian and the car in this accident. The proposed intention prediction model is used to recognize and predict pedestrian intention. The obtained results are shown in Figure 3.21, where the pedestrian initially stopped at the roadside, then crossed the road at $T = 2$, the pedestrian collided with the car around $T = 9$. Based on the probability prediction curve, the proposed method can predict the starting intention of the pedestrian with a higher degree of confidence at $T = 3$. Meanwhile, the proposed system would generate a warning signal for the driver. As the predicted probability of crossing has reached 0.9 at $T = 4.5$, the driver should be aware of the crossing intention of the pedestrian and will slow down the vehicle. Based on the car speed, the longitudinal distance between the car and the pedestrian

Table 3.6: Explanation of parameters related to the accident scene

Parameter	Value
Pedestrian speed V_p	8 km/h
The initial distance between the pedestrian and the collision location D_p	18 m
The initial distance between the pedestrian and the collision location D_c	18.5 m
Vehicle speed V_c	45 km/h
Lateral distance from the collision location to the right side of the road S	13.71 m

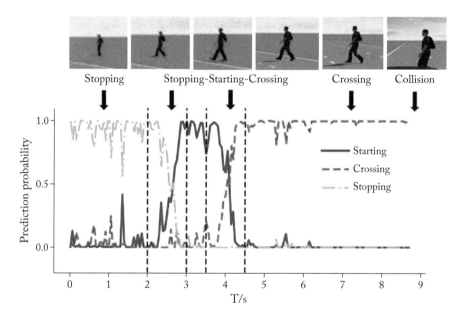

Figure 3.21: The probability of predicted intention of the pedestrian.

is about 56 m at $T = 4.5$, while the braking distance is only 18.5 m. Thus, the car has enough safe distance to avoid a collision with the pedestrian. Thus, the proposed method in this chapter could timely effectively predict the movement of pedestrians, providing vital information for drivers' decision-making, such that, to improve the safety of drivers and pedestrians, and reduce collision accidents.

3.6   SUMMARY

This chapter proposed an LSTM-based framework of pedestrian intention and trajectory prediction, which are experimentally analyzed and verified based on the Daimler dataset. Predicting pedestrian movement can provide useful information for vehicle decision-making, reducing accidents, and improving the safety of vehicles and pedestrians, the obtained results show that:

1. The introduction of head orientation features can enhance the representation of pedestrian motion features and improve the accuracy of intention prediction. It also can improve the ability to predict bending intentions in advance when dealing with pedestrian intention changes, such that it may offer more time for the decision-making of intelligent vehicles.

2. The performance of the trajectory prediction method based on the LSTM network is better than the prediction method based on the kinematics model. The introduction of the attention mechanism for the LSTM network can improve the ability to extract historical trajectory features, thereby improving the performance of the network.

3. The fusion of pedestrian intentions can improve the prediction accuracy of the pedestrian trajectory prediction network, especially when dealing with intention changes. Also, this method can reduce the uncertainty trajectory and the prediction errors of pedestrian movement effectively.

CHAPTER 4

Driver Secondary Driving Task Behavior Recognition

In this chapter, we present a driver activity recognition framework that can steadily extract the driver skeleton data from low-cost monocular cameras, while also, accurately recognizing different driving-related tasks in real time. However, several difficulties should be overcome to achieve this goal. First, the videos collected by the onboard monocular camera always contain noises because of non-negligible environmental factors such as illumination and camera motions, which will directly affect skeleton joints detection. Second, the driver activity recognition network should be of a concise framework with accurate classification performances since the computing resource on-board is usually limited. Third, there usually exists a huge data imbalance among normal driving and other secondary driving tasks in real-time applications, and this will largely affect the recognition performance of small-scale categories.

To overcome those difficulties, an end-to-end driver activity recognition system, namely, the Spatial-Temporal Graph Convolutional LSTM (ST-GCLSTM) networks is proposed. The whole architecture is depicted in Figure 4.1. Positions of the driver's upper body joints are first captured in original frames. Aiming to correct the lost or misrecognized joints, the temporal exponential mean filter is utilized to smooth skeleton data through the sliding window. To formulate the discriminative spatial-temporal feature, the concise five-layers of GCN are proposed to build the driver skeleton graph, meanwhile, the spatial structure representation among joints is derived. Then, spatial features captured by the GCN are transferred into a single layer of attention-enhanced LSTM networks for temporal motion feature extractions among consecutive frames. To solve the data imbalance problem, the Focal loss function is utilized to balance the loss value among the normal driving and other driving-related tasks, eventually, a reasonably structured dataset is collected for model training and evaluations.

The remaining contents of this chapter are reminded as follows. Section 4.2 introduces the design of the driver behavior dataset, and a large-scale driver behavior video dataset composed of natural driving data and simulated driving data is obtained. The detailed modeling of the ST-GCLSTM, the related experiments, and the model evaluations are interpreted in Section 4.3. Finally, conclusions are presented in Section 4.4.

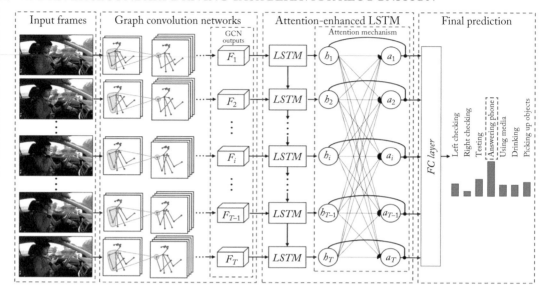

Figure 4.1: The architecture of spatial-temporal graph convolutional LSTM networks.

4.1   DRIVER BEHAVIOR DATASET DESIGN

In this section, the data collection procedure, the relevant data processing method, and the dataset structure will be introduced.

4.1.1   DATA COLLECTION PROCEDURE

The existing datasets related to driver behavior recognition are mainly Kaggle State Farm Distracted Driver Detection (SFDDD) datasets. Kaggle is a platform related to machine learning and deep learning which hosts large databases and provides developers with competition for code writing, sharing, and communication. Kaggle released the driver distraction dataset for State Farm in 2016. The SFDDD dataset was collected by 26 volunteers of different skin colors and different genders. The datasets consist of 10 driving-related tasks, including normal driving, right-hand texting, right-hand calling, left-hand texting, left-hand calling, using the center console, drinking, fetching items inside the vehicle while driving, makeup, and talking to passengers. However, the SFDDD dataset only contains the single frame image of every specific task, and it loses the continuous picture sequences of driver behavior. In addition, due to related policy restrictions, this dataset is only allowed to be used in Kaggle competitions. Considering that human behavior is a continuous process in the temporal domain, and we expect to extract the temporal feature of human behavior from video sequences. Therefore, in this chapter, we design the specific data collection process and construct the driver behavior video dataset.

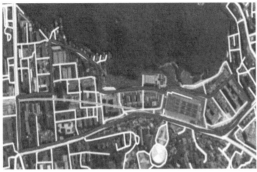

(a) Naturalistic driving route   (b) The test vehicle

Figure 4.2: The naturalistic driving data collection experiment.

We set a low-cost monocular camera with a resolution of 1080P on the right-A-pillar of the vehicle. The original video resolution is 1920×1080. To reduce the computation cost, the video resolution is compressed to 320×180, which will not influence the recognition performance. A total of 7 driving volunteers were recruited in the data collection process, including 6 male drivers and 1 female driver. The average age of driving volunteers was 23 years and the average driving age was 3.9 years. Referring to the Kaggle dataset and the SHRP2 natural driving database, we define 8 common driving behaviors, namely the *normal driving, left or right checking, texting, answering the phone, using media, drinking, and picking up objects,* where the first three tasks are normal driving behaviors, while the last five tasks are distracted driving behaviors. The data collection process can be briefly divided into two steps, including natural driving data collection and simulated driving data collection.

(1) *Naturalistic driving data collection*:   For the sake of security, we select a routine inside the campus for natural driving dataset collection, as shown in Figure 4.2a. The traffic scene is relatively simple, with fewer pedestrians and vehicles. The test vehicle is shown in Figure 4.2b. Each volunteer performs the above eight driving behaviors naturally adapting to their driving habits, and the order of execution between different driving behaviors is random. In addition, to ensure the safety of the data collection process, a director will monitor the behavior of the volunteers and the current road conditions at all times, and promptly give guidance or warnings when necessary.

(2) *Simulated driving data collection*:   Because the number of distracted driving samples in the naturalistic driving data is very limited, it cannot meet the training requirements of deep neural networks. Thus, the volunteers were asked to mimic the driving-related tasks in a stopped vehicle. Again, volunteers would be required to perform secondary tasks naturally adapting to their driving habits and experience. In addition, this experiment requires volunteers to resume normal

Table 4.1: Illustration of the obtained datasets

Driving Task	Size of Video Sequences
Normal driving	5012
Left checking	894
Right checking	1066
Texting	1374
Answering cell phone	1158
Using media	1530
Drinking	839
Picking up	675

driving for 1 min every time they finished the distracted driving behavior and the execution order is random so that the simulative data of driving-related tasks would be more representative of naturalistic driving conditions.

Each volunteer needs to go through the experimental data collection process for about 90 min. Then the video data will be cut into corresponding clips according to the above-mentioned driver behavior categories. Most driving behaviors last only a few seconds, while some faster driving behaviors, such as *checking the left* or *right rearview mirror*, last varying from 0.5–1 s. Meanwhile, the input of the proposed algorithm needs a fixed-length video sequence. To balance the temporal variations among different driving behaviors, we set the sequence length to 15 frames, that is, the duration of every video sequence is 1 s, and apply the sliding window method to process the driver behavior videos into fixed-length video sequences. Finally, the total number of video sequences is 12,177. The structure of the dataset and the number of video clips of every driving behavior is shown in Table 4.1.

4.1.2   DATA PREPROCESSING

At first, we will use the AlphaPose algorithm to extract the driver's key point. The AlphaPose algorithm uses the COCO dataset for training, so the original output of the AlphaPose algorithm contains a total of 18 key points, and the output format of each key point is represented by (x, y, c), where x and y denote the pixel coordinates of the key point, and c is the confidence of the coordinate point category. The human body keypoints indexes of the original output are shown in Figure 4.3.

However, in the naturalistic driving conditions, the illumination varies and the lower body of the driver might be missing in the camera view, the adaptability of the original AlphaPose could be underperforming in this condition. Therefore, we use the COCO datasets and our

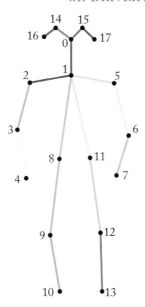

Figure 4.3: Original COCO dataset annotation format.

collected datasets to fine-tune the last few layers of the pre-trained model. Specifically, we replace the model with a new fully connected layer that only outputs the 14 joints of the driver's upper body. Considering that the secondary task engagement while driving is a more complex indoor scenario due to the illumination and the camera's shake, some joints of the human skeleton would be lost or misrecognized. Thus, we use the temporal exponential mean filter and set a confidence threshold to correct the misrecognized joints. Specifically, we set a temporal sliding window of five frames and if the confidence value of the joint is below the threshold, the joint will be replaced by the exponential mean value of corresponding joints in the past five frames. The function of the temporal exponential mean filter is defined in Equation (4.1).

$$\overline{n}_t^i = \begin{cases} n_t^i, \ conf\left(n_t^i\right) \geq 0.8 \\ \alpha \sum_{m=0}^{5} (1-\alpha)^m \, n_{t-m}^i, \ \text{otherwise}, \end{cases} \tag{4.1}$$

where α denotes a hyper-parameter which is set to 0.7.

In the temporal exponential average filter, the higher weight is given to the closing frame, which is consistent with the actual situation. All skeleton joints' position would be filtered, one example of the smooth result regarding the driver's right wrist is shown in Figure 4.4. It can be seen that with the temporal exponential mean filter, the fluctuation of the skeleton key point is greatly reduced, thereby reducing the influence of the skeleton key point data noise on the recognition accuracy of the model.

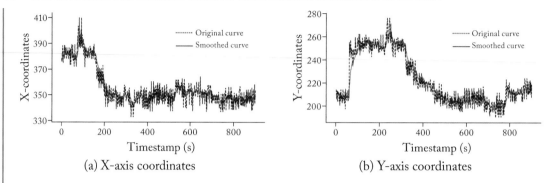

Figure 4.4: Coordinates smoothing of right wrist.

4.2   DRIVER ACTIVITY RECOGNITION USING SPATIAL-TEMPORAL GRAPH CONVOLUTIONAL LSTM NETWORK

In this section, the algorithm of the proposed ST-GCLSTM model is introduced in detail. More specifically, the process of graph construction, the implementation of graph convolution, the construction of attention-enhanced LSTM networks are presented.

4.2.1   SPATIAL-TEMPORAL GRAPH CONVOLUTIONAL LSTM NETWORKS

As introduced above, we first use the pre-trained AlphaPose algorithm to extract the pixel coordinates of the key points of the driver's skeleton from the video image. Let $N_t = \{n_t^1, n_t^2, \ldots, n_t^m\}$ denotes the key point of the driver's skeleton at time t, where $n_t^i = (x_t^i, y_t^i)$, x_t^i, y_t^i are the pixel x-coordinate and y-coordinate of the key point of the driver skeleton, respectively, m is the number of key points, which is $m = 14$ in this study.

However, the original outputs of the AlphaPose algorithm are the only concatenation of the skeleton key points in a certain order, and there is no connection relationship between the key points. In fact, the human body skeleton has its spatial structure, and each key point should have a corresponding connection relationship based on the human body structure. In addition, human body movements are usually completed by the cooperation of multiple limbs. For example, in the action of picking up objects, there is a process of "positioning" and "holding." The human body first "positions" the target object through the eyes, and then performs the "holding" through the hands. Although there is no physical connection between the hand and the eye, there is a certain cooperative relationship between the two limbs. Therefore, to preserve the interaction between the key points, we define the key point connectivity set $L_t = \{n_t^i n_t^j | n_t^i, n_t^j \in N_t, \overline{n_t^i, n_t^j}\}$, where n_t^i, n_t^j indicates that the key point n_t^i is connected to the point n_t^j. For matrix operation, L_t can

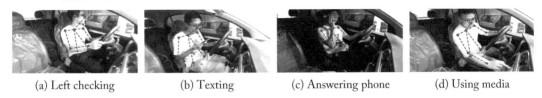

| (a) Left checking | (b) Texting | (c) Answering phone | (d) Using media |

Figure 4.5: Illustration of raw images and skeleton estimation.

also be denoted as the adjacency matrix $A_t \in \mathbb{R}^{m \times m}$, as shown in Equation (4.2):

$$A_t(i, j) = \begin{cases} 1, & n_t^i n_t^j \in L_t \\ 0, & \text{otherwise.} \end{cases} \tag{4.2}$$

Concatenate the key point set N_t and the adjacency matrix A_t, we can get the skeleton graph of the single-frame at time t, $G_t = \{N_t, A_t\}$. And the input sequence of skeleton graphs is denoted as $I = \{G_1, G_2, \ldots, G_T\}$, where T is the length of the sequence. The skeleton estimation results of the video images are shown in Figure 4.5.

In the traditional convolution operation, the pixels of the image are arranged in a European spatial structure from left to right and top to bottom, so there is a natural traversal order. In addition, after a given convolution kernel size and sliding step size, the number of pixels in the receptive field is also fixed, so traditional convolution operations can achieve weight sharing, which greatly reduces the number of network parameters. Different from normal images, the key points of the skeleton graph are in non-Euclidean space, and the key points have no fixed traversal order or mutual arrangement structure. Therefore, the sampling function and weight function should be rebuilt in graph convolution.

In the traditional convolution operation, the sampling function uses distance as the sampling standard. Similarly, the graph sampling function uses the step length between key points as the sampling standard, where the key point step length means the shortest path length between the root key point and the sampling point in the skeleton graph. Therefore, the graph sampling function is defined as Equation (4.3):

$$S\left(n_t^i, D\right) = \left\{n_t^j | d_{ij} \leq D\right\}, \tag{4.3}$$

where D denotes the sampling step between nodes. Let $D = 1$, which means the sampling points are directly connected to the root key points.

We will further simplify the process of nodes labeling process and weight distribution. The graph connectivity could be represented by a connectivity set L_t, therefore, the labeling process can be simplified by manually defining the order of all nodes, as shown in Figure 4.6b. In traditional convolution, given the specific convolution kernel, the size of the receptive field is fixed so that the convolution kernel could be shared on the input feature, which could largely

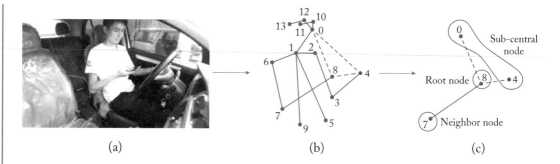

(a)   (b)   (c)

Figure 4.6: The process of the partition strategy. (a) Raw image of the texting task. (b) Skeleton graph and nodes labeling. (c) Receptive field selection and nodes partition.

reduce the parameter amount of the network. However, in graph convolution, the node number of receptive fields varies with the different root joints. If every key point is assigned with a unique weight, the complexity of the convolution operation will be greatly increased. In this chapter, we propose a part-aware partition strategy, as shown in Figure 4.6c.

Generally, the human body moves in local parts, which means, eyes, head, and hands will corporately sustain the task load when pointing to a specific visual target. We define the key points No. 0, No. 4, and No. 8 as the sub-central center of the head, left hand, and right hand, respectively. The connection of the whole skeleton graph is shown in Figure 4.6b. Then, the key point in the receptive field would be divided into three subsets: (i) root nodes, (ii) sub-central nodes (the center of limbs), and (iii) neighbor nodes (remaining nodes in the receptive field). The process of partition strategy could be denoted as Equation (4.4):

$$P\left(n_t^i, n_t^j\right) = \begin{cases} 1, & n_t^j = n_t^i \\ 2, & n_t^j \in \{n_t^0, n_t^4, n_t^8\} \\ 3, & \text{otherwise.} \end{cases} \tag{4.4}$$

Every subset would share a trainable weight, thus the graph convolution kernel size could be fixed to 3×1. Note that if there are not enough joints to be divided into three subsets, the zero-padding strategy will be adopted to fit the vacant subset. In this way, the weight sharing mechanism can also be implemented in the graph convolution, which greatly simplifies the complexity of the graph convolution operation and reduces the parameter amount of the network.

Combining Equation (4.3) and (4.4), the graph weight function of the root joints n_t^i at time t could be denoted as Equation (4.5):

$$W_{n_t^i}\left(n_t^j, c\right) = W\left(P\left(n_t^i, n_t^j\right), c\right), \; n_t^j \in S\left(n_t^i, D\right), \tag{4.5}$$

where c denotes the channel number of the feature map.

According to the graph sampling function, the graph weight function, and the partition strategy introduced above, the output feature vector $\boldsymbol{F}_{out}^t$ of the node n_t^i at the time step t is defined as Equation (4.6):

$$F_{out}^t \left(n_t^i \right) = \sum_{c=0}^{C} \sum_{n_t^j \in S \left(n_t^i, D \right)} F_{in}^t \left(n_t^i, n_t^j, c \right) \boldsymbol{W} \left(\boldsymbol{P} \left(n_t^i, n_t^j \right), c \right), \tag{4.6}$$

where $\boldsymbol{F}_{out}^t$ denotes the output feature vector, $\boldsymbol{F}_{in}^t$ denotes the input feature vector, $\boldsymbol{W}()$ denotes the weight distribution function, $\boldsymbol{P}(\cdot)$ denotes the partition strategy, and $\boldsymbol{S}()$ denotes the sampling function.

As mentioned above, the joints connectivity can also be represented by the adjacency matrix $\boldsymbol{A}_t$. According to the joints partition strategy, the adjacency matrix can be divided into three sub-matrices, as shown in Equation (4.7):

$$\boldsymbol{A}_t = \sum_l \boldsymbol{A}_t^l, l \in \{1, 2, 3\}, \tag{4.7}$$

where $\boldsymbol{A}_t^1$ contains the self-connection of root nodes, $\boldsymbol{A}_t^2$ contains the connection between root nodes and sub-centers, and $\boldsymbol{A}_t^3$ contains the connection between root nodes and neighbor nodes. Therefore, the graph convolution operation on the entire skeleton graph could be implemented as Equation (4.8):

$$F_{out}^t = \sum_{c=1}^{C} \sum_{l=1}^{3} Z_t^{l-\frac{1}{2}} A_t^l Z_t^{l-\frac{1}{2}} F_{in}^t(c) \boldsymbol{W}(l, c), \tag{4.8}$$

where $\boldsymbol{Z}_t^l$ is a normalized matrix, which is used to balance the contribution of the different partitions. The calculation of the $\boldsymbol{Z}_t^l$ is denoted as Equation (4.9):

$$Z_t^l(i, i) = \sum_j A_t^l(i, j). \tag{4.9}$$

We construct a concise and lightweight graph convolutional network, which contains in total of five graph convolutional layers. The architecture of the graph convolutional networks is shown in Table 4.2. By sending each frame skeleton graph into the graph convolutional networks, the human skeleton spatial structure feature vector $\boldsymbol{F}_{out}^t$ could be obtained at each moment, and the human skeleton spatial structure feature sequence could be denoted as $V = \{\boldsymbol{F}_{out}^1, \boldsymbol{F}_{out}^2, \ldots, \boldsymbol{F}_{out}^T\}$, where T is the sequence length.

4.2.2   TEMPORAL LSTM

To construct the relation between the spatial configuration and the temporal dynamics, the LSTM networks are adopted on top of the aforementioned GCNs. The spatial structure feature $V = \{\boldsymbol{F}_{out}^1, \boldsymbol{F}_{out}^2, \ldots, \boldsymbol{F}_{out}^T\}$ captured from the GCN is transferred into the LSTM networks.

Table 4.2: Architecture of the GCN

Graph Convolutional Networks	
Input	$I = \{G_1, G_2, \ldots, G_T\}$
Graph convolution layers	3×1 G-Conv 96
	3×1 G-Conv 256
	3×1 G-Conv 384
	3×1 G-Conv 384
	3×1 G-Conv 256
	Average pooling
Output	$V = \{F_{out}^1, F_{out}^2, \ldots, F_{out}^T\}$

Through the input gate i_t, forget gate f_t and output gate o_t, the LSTM cells can learn how much information to store or discard and update the hidden state h_t timely according to the input data v_t and h_{t-1}. For the first of the LSTM layer, $v_t = F_{out}^t$. The functions of the basic LSTM cell are defined as Equation (4.10):

$$\begin{cases} f_t = \sigma\left(W_f\left[h_{t-1}, v_t\right] + b_f\right) \\ i_t = \sigma\left(W_i\left[h_{t-1}, v_t\right] + b_i\right) \\ u_t = \tanh\left(W_u\left[h_{t-1}, v_t\right] + b_u\right) \\ o_t = \sigma\left(W_o\left[h_{t-1}, v_t\right] + b_o\right) \\ c_t = i_t \odot u_t + f_t \odot c_{t-1} \\ h_t = \tanh\left(o_t\right) \odot c_t, \end{cases} \tag{4.10}$$

where $\odot$ denotes the element-wise product, and σ represents the sigmoid function.

Generally, the hidden state of the last memory cell of the LSTM layer, i.e., h_T, is adopted as the final output vector for classification. However, human action is an evolutionary process, so contributions are different among different frames. For example, when the driver is answering the phone, the frames of holding the phone near the ear are certainly more representative than those frames of lifting the phone. To fully utilize all hidden states of the LSTM memory cells in the last layer and adaptively emphasize key frames, the attention mechanism on the top of the LSTM layers is utilized to allocate the trainable weight value for every hidden state. Functions of the attention mechanism are constructed as Equation (4.11):

$$\begin{cases} M = \tanh\left(H\right) \\ \alpha = \text{softmax}\left(w^T M\right) \\ H^* = H\alpha^T, \end{cases} \tag{4.11}$$

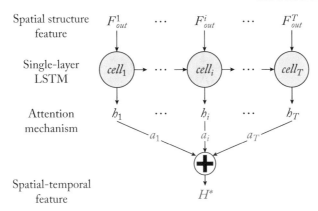

Figure 4.7: The structure of the attention-enhanced LSTM networks.

where $\boldsymbol{H}_{n \times T}$ is the output feature vectors of LSTM layers, n denotes the size of each LSTM memory cell, T denotes the length of the sequence, $\boldsymbol{w}$ denotes the trainable weight vector, and $\boldsymbol{\alpha}$ denotes the distribution coefficient matrix. Dimensions of $\boldsymbol{w}$, $\boldsymbol{\alpha}$ are n and T, respectively. $\boldsymbol{H}_{n \times T}^{*}$ refers to the final feature vector of the attention model. Figure 4.7 shows the structure of the attention-enhanced LSTM networks. The spatial feature captured by the GCN is transferred into a single-layer LSTM network. Each memory cell calculates spatial-temporal features through the state transition. While the attention mechanism allocates each memory cell a unique weight and captures the synthetically spatial-temporal feature of the ST-GCLSTM model. Finally, the fully connected layer with the SoftMax classifier could compute the score of each class from the spatial-temporal feature of drivers' activity.

4.3   MODEL EVALUATION

Detailed descriptions of the solution of data imbalance will be introduced in this subsection first. Then, the ablation study and the comparative study with other methods are conducted to evaluate the effectiveness of the proposed model. Finally, the real-time feasibility of the proposed architecture is evaluated, verified, and discussed.

(1) *Data imbalance:*   In the naturalistic driving condition, the driver is in a normal driving state most of the time. Compared to the normal driving task, the number of the second driving task is very limited. Even though we utilize the simulated driving process to enlarge the number of secondary tasks and down-sampling the normal driving data, the problem of data imbalance between the normal driving task and the second driving task is still very serious. In the training process of deep neural networks, the problem of data imbalance will lead the network to tend to emphasize the categories with large numbers while ignoring categories with a small number, which will seriously affect the recognition accuracy of small-number categories. Therefore, Focal

Loss is used as the loss function for network training. The equation of Focal Loss is shown in Equation (4.12):

$$L_{FL} = -\frac{1}{N} \sum_{i}^{N} \sum_{c=1}^{M} \alpha_c \left(1 - p_{i,c}\right)^{\gamma} y_{i,c} \log\left(p_{i,c}\right), \tag{4.12}$$

where N denotes the number of samples, M denotes the number of classes, $y_{i,c}$ denotes the binary indicator (0 or 1), $y_{i,c} = 1$ if class label c is the correct classification for observation i, and $p_{i,c}$ is the predicted probability of class c for observation i. α_c is a balance factor, and we set different α_c for every class so that the loss of small-scale class will be increased and the loss of large-scale class will be reduced. γ is a modular factor that could increase the loss of hard samples while reducing the loss of easy samples to improve the robustness of the model. According to experiments, $\alpha_c = 0.7$ if class label c is normal driving, otherwise $\alpha_c = 0.3$, and $\gamma = 2$.

(2) *Evaluations of ST-GCLSTM Model*:   The effectiveness of the attention mechanism module and the graph convolution network will be evaluated through an ablation study. The baseline model of this chapter is a single-layer LSTM network with 128 units whose input is the simple concatenation of the skeleton joints position, namely, referring to the Basic-LSTM. Since the recall ratio means the fraction of the total amount of positive instances that were recognized, for the sake of safety, a high recall ratio of secondary task engagements indicates a safe and concentrating driving maneuver, instead, accuracy and precision are inadvisable in data imbalance situation. Thus, we mainly focused on the recall ratio in evaluation indexes.

First, the effectiveness of the graph convolution network is evaluated. We add the graph convolutional network based on the Basic-LSTM network, namely, the ST-GCLSTM network (no attention). Figures 4.8 and 4.9 show the confusion matrix of the Basic LSTM network and the ST-GCLSTM network, respectively. The first column on the right side reflects the recall ratio and the bottom row shows the precision ratio. The overall recall rates of the Basic-LSTM network and ST-GCLSTM network are 82.1% and 87.9%, respectively. It can be seen that the improvement effect of the graph convolutional network is very significant.

In addition, the ST-GCLSTM network has improved the recall rate of every task to varying degrees. Regarding the Basic-LSTM network, the biggest confusion in secondary tasks exists between drinking and answering cell phones. Almost 22.2% of drinking cases were incorrectly identified as answering cell phones. In the ST-GCLSTM network, the confusion between these two tasks is greatly reduced, and the recall rate of the drinking task is increased by 8.9%. In addition, texting has achieved the most significant improvement, and the recall rate of texting has increased by 12.3% compared to the baseline. At the same time, the confusion between the normal driving and the second driving tasks in the SG-GCLSTM network is greatly reduced. The misrecognition ratio between normal driving and texting is reduced by 13.4%. Compared with the Basic-LSTM network, the ST-GCLSTM network establishes the connectivity between key points and extracts the subtle spatial structure features of the driver's skeleton through the multi-layer graph convolution. The facts above reflect that the complicated

	T1	T2	T3	T4	T5	T6	T7	T8	
T1	2536	848	156	1436	8	24	4	0	50.6%
T2	55	778	14	45	0	2	0	0	87.0%
T3	11	4	983	42	0	26	0	0	92.2%
T4	14	73	19	1140	69	27	8	24	82.9%
T5	0	3	8	96	1009	0	30	4	87.1%
T6	7	3	13	0	0	1136	0	0	98.0%
T7	51	22	11	6	186	40	523	0	62.3%
T8	0	0	8	12	0	0	0	655	97.0%
	94.8%	46.8%	81.1%	41.1%	79.3%	90.5%	92.5%	95.9%	82.1%

Figure 4.8: The confusion matrix of the Basic-LSTM network.

	T1	T2	T3	T4	T5	T6	T7	T8	
T1	3208	828	72	648	172	52	32	0	64.0%
T2	42	802	0	50	0	0	1	0	89.7%
T3	11	0	1043	7	0	5	0	0	97.8%
T4	16	23	7	1309	10	3	0	6	95.2%
T5	0	2	0	137	1002	0	17	0	86.5%
T6	6	1	0	2	0	1150	0	0	99.2%
T7	28	7	57	14	95	40	598	0	71.2%
T8	0	0	2	0	0	0	1	672	99.5%
	96.9%	48.2%	88.2%	60.4%	78.3%	92.0%	92.1%	99.1%	87.9%

Figure 4.9: The confusion matrix of the ST-GCLSTM network.

processing of the graph construction extracts effective spatial features among human skeleton positions and improves the diver activity recognition performance.

Then we compare the ST-GCLSTM and the ST-GCLSTM with attention to evaluate the influence of the attention mechanism. Figure 4.10 shows the confusion matrix of the ST-GCLSM with attention. The general recall ratio is 88.8%, which is slightly higher than the

	T1	T2	T3	T4	T5	T6	T7	T8	
T1	3636	756	84	524	8	0	4	0	72.5%
T2	6	831	10	26	6	0	0	15	92.9%
T3	28	4	1024	10	0	0	0	0	96.1%
T4	13	81	5	1186	69	0	4	16	86.3%
T5	1	19	0	4	1130	0	4	0	97.5%
T6	5	0	4	3	0	1103	40	4	95.1%
T7	3	40	0	7	149	40	600	0	71.5%
T8	0	1	0	7	0	0	0	667	98.8%
	98.4%	47.9%	90.8%	67.1%	82.9%	96.5%	92.0%	95.0%	88.8%

Figure 4.10: The confusion matrix of the ST-GCLSTM-Attention network.

Table 4.3: Classification result of ST-GCLSTM with attention model using the LOO valuation

	T1	T2	T3	T4	T5	T6	T7	T8	Mean
D1	0.267	1.0	0.922	0.989	0.936	0.956	0.592	1.0	0.832
D2	1.0	1.0	1.0	0.981	0.999	1.0	0.711	0.953	0.955
D3	0.965	0.712	0.984	0.795	0.970	0.731	0.857	1.0	0.956
D4	0.529	0.987	0.872	0.763	1.0	0.978	0.667	0.991	0.848
D5	1.0	1.0	0.974	0.986	1.0	0.966	0.573	1.0	0.937
D6	0.409	0.667	0.891	0.526	1.0	1.0	0.922	1.0	0.802
D7	0.977	0.810	1.0	0.477	0.985	1.0	0.916	0.987	0.894
Average	0.725	0.929	0.961	0.863	0.975	0.951	0.715	0.988	0.888

Tips: T1–T8 indicate eight driving-related tasks, while D1–D7 refer to seven different drivers.

ST-GCLSTM. Regarding the task of answering the phone, the ST-GCLSTM with attention gets the improvement by 11%. Considering the answering phone is a long-term activity where its sequence in time-window contains many irrelevant frames. Thanks to the introduced attention mechanism which adaptively allocates every frame in series with a unique weight value, such that critical frames which contain more informative features could be stressed, and the extraction of temporal dependency could be further strengthened.

Table 4.4: Evaluation of the model components

Methods	Precision	Recall	F1 Score
Basic LSTM	77.4%	82.1%	79.7%
ST-GCLSTM	81.9%	87.9%	84.7%
ST-GCLSTM with Attention	83.8%	88.8 %	86.2%

Table 4.3 shows the cross-validation classification results of the ST-GCLSTM-Attention network. The leave one out (LOO) cross-valuation method is utilized which could better verify the extensiveness of the proposed model. Specifically, the data of arbitrary one driver is used as the testing dataset and the rest is used as the training dataset. Therefore, the testing data of each driver are new to the model. The first column on the right side indicates the mean recall ratio of every driver toward all eight driving tasks, while the bottom row depicts the average recognition result of each driving task among all drivers. Specifically, the classification recall ratios of all drivers are higher than 80% and Driver #3 gets the highest 95.6% recognition recall ratio. The worst classification performance (80.2%) is found in Driver #6. One reason is that Driver #6 who is a cautious female driver tends to perform driving tasks slightly and swiftly where the duration is too short to capture enough motion features. Moreover, Driver #6 tends to use eye movement more than head movement to check targets, and those body dynamics are harder to extract. Regarding the result of secondary tasks, picking up gets the best recall ratio of 98.8%, while drinking gets the lowest recall ratio of 71.5%. The results are all consistent with the confusion matrixes as shown in Figure 4.10.

Table 4.4 further shows the recognition macro precision ratio, recall ratio, and F1 score of the aforementioned models in the study of component comparisons. Figure 4.11 shows the PR (Precision-Recall) curves of the baselines and the proposed method. It can be seen that the ST-GCLSTM-Attention network has achieved the best recognition performance, and the implementations of the graph convolution and attention mechanism both improve the performance of the baseline in varying degrees.

4.4   COMPARATIVE STUDY AND REAL-TIME APPLICATION

To further reflect the effectiveness of the proposed model, we also compare the proposed method with other relevant methods. Please note that the reproduction result of other methods may be slightly lower than that of the original ones, because the experiment uses the collected dataset by ourselves to ensure data consistency, and uses the leave-one-out method for cross-validation, which is stricter than the normal cross valuation method. In general, the selected related works in the comparison could be divided into three typical types.

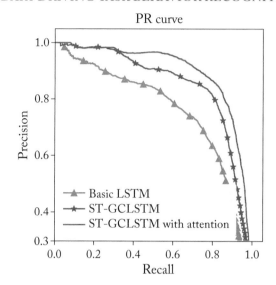

Figure 4.11: Comparison of PR curves between the proposed method and baselines.

The biggest difference between the ST-GCLSTM-Attention network and those variations of the LSTM network (Geometric-LSTM network) lies in the input features. Existing
variations of the LSTM network usually use discrete skeleton key point sets, or manual geometric features, such as distance, angle, etc. The proposed model uses the graph convolutional
network to construct the human skeleton graph, establishes key point connectivity, and adaptively extracts skeleton spatial structure features, so the input features will be more succinct and
informative in semantic information. Through the unique graph convolution operation, the proposed model substantially improves the performance of LSTM.

(1) *Graph-convolution-based method*:   The ST-GCN network also constructs the human
skeleton graph. In addition, The ST-GCN constructs both spatial and temporal connectivity
among sequential skeletons. Compared with the ST-GCN network, the model in this chapter
uses LSTM for temporal feature extraction, which can fuse the information of different frames
and different key points to obtain comprehensive temporal and spatial features. Instead, the ST-
GCN network only computes the temporal convolution on the same joints between consecutive
frames.

(2) *Convolutional neural network-based method*:   GMM-AlexNet implements end-to-end
driver behavior classification based on the original single frame image, but the model ignores
the temporal relationship between consecutive frames. Compared with the GMM-AlexNet,
the proposed model has carried out sufficient feature extraction in both spatial and temporal

Table 4.5: Comparisons with other methods

Methods	Precision	Recall	F1 score
Geometric-LSTM [91]	77.0%	85.0%	80.8%
ST-GCN [92]	80.1%	86.5%	83.1%
GMM-AlexNet [178]	81.9%	87.4%	84.5%
ST-GCLSTM with Attention (Proposed)	83.8 %	88.8%	86.2%

dimensions. In addition, the use of the human skeleton as a representation of the human body structure also greatly reduces the interference of background noise.

Table 4.5 shows the comparison of the classification results between the ST-GCLSTM-Attention network and the above-mentioned existing models. Figure 4.12 shows the comparison of the PR curves of each model. From Table 4.5 and Figure 4.12, it can be seen that the ST-GCLSTM-Attention network is superior to the existing models in the accuracy of driver behavior recognition, and has achieved excellent robustness and generalization.

To further verify the feasibility of the implementation of the ST-GCLSTM-Attention network, this chapter designs the real-time application experiments. The test platform used in this chapter is a laptop with NVIDIA GTX1060, and the operating system is Linux. Image preprocessing and AlphaPose's calculation of the driver's skeleton key points from the original image takes about 55 ms, while the ST-GCLSTM-Attention network takes only 9 ms to recognize the driver's behavior from the skeleton key points, so the real-time application calculation efficiency of ST-GCLSTM-Attention is about 15 fps. The real-time application experiment in this chapter only runs on a PC, and the programming language is Python. In addition, the reasoning framework of the current model is PyTorch. Since the model is not pruned and quantified, the reasoning speed still needs to be improved. On embedded platforms with limited computing power, the TensorRT deep learning reasoning framework can be used to increase the speed of reasoning. At the same time, the image preprocessing and model output post-processing processes can be written in C++ language to further improve operating efficiency.

Figure 4.13 shows the confidence curves of different driving-related activities in real-time applications, which demonstrates the process of the transformation from normal driving to texting. The lines and shades reflect the mean probability and the standard deviation of different tasks, and the two blue dashed lines reflect the changing point of different tasks. The proposed ST-GCLSTM model could effectively and precisely recognize driver activity. In future work, the datasets will be enriched and more categories will be included so that the real-time application could be further improved.

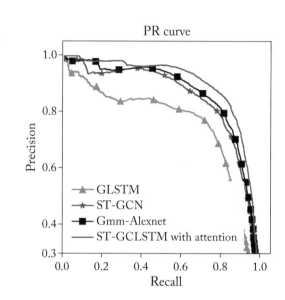

Figure 4.12: Comparison of the PR curves between the proposed method and other methods.

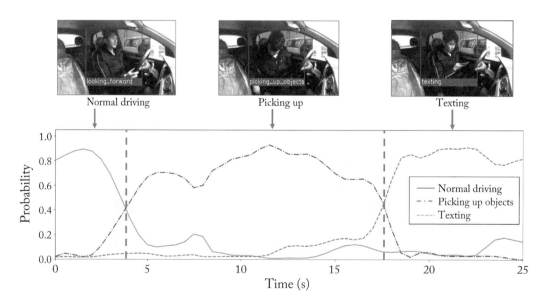

Figure 4.13: Results of driver activity recognition in real time.

4.5   SUMMARY

A driver activity recognition system is proposed in this chapter, in which the human body skeleton data is extracted from raw videos to represent human body motion. To reduce the noise of raw joints data, the temporal exponential mean filter is utilized to correct the lost or wrong predicted joints. Then, the ST-GCLSTM model with attention mechanism is constructed to reason spatial features among different skeleton joints as well as capture temporal dependency through consecutive frames. To overcome the data imbalance problem, a well-structured driving task dataset is collected, and the Focal loss function is utilized for model training. Manifold comparisons of experiments show that the proposed ST-GCLSTM model achieves an 88.80% recall ratio of 8 common driving-related tasks. The real-time application efficiency reaches up to 15 fps, which has the potential to transfer to embedded hardware and satisfy the engineering application.

CHAPTER 5

Car-Following Driving Style Classification

In this chapter, we propose a driving style classification method based on the car-following event extracted from the Next Generation Simulation (NGSIM) dataset that combines the Principal Component Analysis (PCA) and Gaussian Mixture Model (GMM). To fully describe the features of the car-following event as possible, we select multivariate parameters which contain the statistical features of the samples from the time and frequency domain. Through the PCA, we choose the main principal component to build comprehensive indicators. The GMM algorithm is applied to classify the aggressiveness of the car-following event through cluster analysis, which is expected to reduce the influence of data distribution on the driving style classification results. Finally, we will talk about the influence of the driving environmental factors (road types and weather conditions) on car-following driving style.

5.1   DATA PREPARATION

Naturalistic driving data are generated by drivers in a daily driving process without interference or the presence of experimenters, which can truly reflect the behavioral characteristics of drivers, this study is based on the naturalistic driving data from the NGSIM database, which is collected by the Federal Highway Administration (FHWA) of the United States [132]. The dataset contains rich vehicle state information related to vehicle dynamics such as longitudinal velocity, longitudinal acceleration, longitudinal position, lateral position, and so on. In addition, Basu et al. [101] found that driver's driving performance might vary due to different environmental conditions. The NGSIM project team collected the driving data on fixed road segments in small periods, while both road conditions and weather change little, which reduces the influence of environmental factors on drivers. Thus, the NGSIM dataset is suitable for conducting driving style-related research.

We extracted the data from the I80 road section in the NGSIM dataset, in which Table 5.1 shows the selected data types. The driving data of the I-80 road section contains the trajectory of all vehicles on the interstate highway collected by the NGSIM project team in the San Francisco Bay Area in Emeryville, CA, on April 13, 2005. The data collection area is about 500 m long, which consists of six freeways. The NGSIM project team records the vehicle information with a sampling frequency of 10 Hz.

Table 5.1: Data types in NGSIM dataset

Data Types	Units and Remarks
Vehicle ID	Integer
Global time	ms
Vehicle class	1-motorcycle, 2-car, 3-truck
Vehicle velocity	ft/s
Vehicle acceleration	ft/s^2
Lane ID	Integer
Preceding vehicle ID	Integer
Space headway	ft
Time headway	s

5.1.1   CAR-FOLLOWING EVENT DATA EXTRACTION

Basic criteria for car-following events were described in a previous study by Zhu et al. [143]

(1) Longitudinal headway < 120 m, eliminating free-flow traffic conditions.

(2) Car-following duration > 15 s, guaranteeing that the car-following period contained enough data for analysis.

(3) Target identification number detected by the radar remains constant, indicating the subject vehicle was preceded by the same lead vehicle (LV).

(4) Lateral distance < 2.5 m, ensuring that the two vehicles were in the same lane.

However, the criteria (3) and (4) cannot be directly applied to this study because it is difficult to calculate the lateral distance of vehicles based on the information in the dataset, and there are also too many vehicle types in the dataset. Thus, we replaced the two criteria as follows:

(3*) The vehicle model should be a passenger car, avoiding other types of vehicles (e.g., trucks) to confuse the driving style identification.

(4*) Both lane ID and the preceding vehicle ID remain constant, indicating that the subject vehicle is preceded by the same lead vehicle and the two vehicles are in the same lane.

5.1.2   DATA SMOOTHING

As the original data is converted from video streams, we need to smooth the extracted data to reduce the disturbances and noises. The Locally Weighted Regression (LOESS) method is

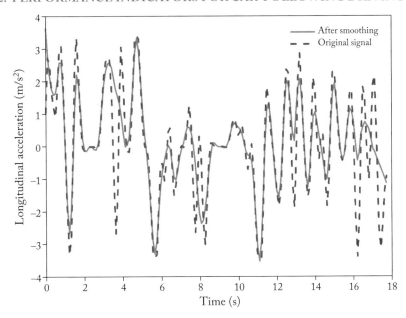

Figure 5.1: The longitudinal acceleration smoothing result of a car-following event from car #936.

one of the commonly used data smoothing methods, which has the advantage of simplicity and efficiency. While the Robust Locally Weighted Regression (RLOESS) is an enhanced version of the LOESS which obtains more robust smoothing results than the LOESS method [144]. Therefore, the RLOESS method is utilized in this study for driving data smoothing, which provides data support for the following study of driving style recognition. For example, the longitudinal acceleration smoothing result of one car-following event from vehicle ID #936 is shown in Figure 5.1. It is easy to verify that the original data fluctuation is more volatile in several segments but less volatile after smoothing.

5.2 PERFORMANCE INDICATORS FOR CAR-FOLLOWING DRIVING STYLES IDENTIFICATION

We intend to select typical features from the time and frequency domains for describing the car-following event based on the previous related studies. They are used to construct comprehensive performance indicators through dimensionality reduction by the PCA approach, eventually, for the car-following driving styles identification.

Table 5.2: Statistical features in the time domain

Time-Domain Parameters	Symbols
Short-time headway proportion	THP_S
The inverse of 95% percentile of the TTC	$\text{TTC}_{inv}^{0.95}$
The standard deviation of acceleration	ACC_{std}

5.2.1   STATISTICAL FEATURES IN THE TIME DOMAIN

Generally speaking, the more aggressive a drivers' driving style is, the more likely the driver is to engage in dangerous driving behavior, thus, the longitudinal acceleration of a vehicle is an effective indicator for identifying the driving style [109]. However, while in a car-following process, variables related to vehicle kinematics such as speed or longitudinal acceleration do not reflect the probability of rear-end collision. Mahmud et al. [145] concluded that time headway (TH) and time to collision (TTC) are important indicators for evaluating the severity of an encounter. The lower the indicators are, the higher the level of a collision will be. Based on the above discussion, we use three time-domain indicators, as shown in Table 5.2, for the driving style classification.

Moreover, the THP_s is calculated by

$$\text{THP}_s = \frac{n_{TH} \leq 1.5\,\text{s}}{N} \times 100, \tag{5.1}$$

where $n_{TH} \leq 1.5$ s denotes the frequency of the data in a car-following event where the *TH* is lower than 1.5 s, and N is the total size of the sample event. If the THP_s of the sample event is high, which indicates the corresponding driving style might be aggressive. In addition, we take the inverse of the TTC instead of the direct value of the TTC because the TTC is distributed in a very large interval that is inconvenient to processes. The event sample with higher TTC_{inv} indicates that the corresponding driving style of this car-following event could be more aggressive than other samples.

5.2.2   STATISTICAL FEATURES IN THE FREQUENCY DOMAIN

Considering that the change of the time headway, acceleration, and speed for the unstable car-following event could be is relatively drastic, to measure the degree of such instability, frequency-domain features have been adopted. More specifically, the power spectral density of the signal is used, which describes how the power of the signal is distributed with frequency and could be defined by

$$p(f) = \lim_{T \to \infty} \frac{1}{T} |S_T(f)|^2, \tag{5.2}$$

Table 5.3: Statistical features in the frequency-domain

Frequency-Domain Parameters	Symbols
The average power of the inverse of the TH	THPow_{inv}
The average power of TTC reciprocal	TTCPow_{inv}
The average power of jerk	JerkPow

where $S_T(f) = S(s(t))$ denotes the Fourier transform of the signal $s(t), t \in [-T/2, T/2]$ and f denote the signal frequency. Then the average power can be further calculated by integrating power spectral density along with the frequency, namely,

$$P = \int_{-\infty}^{+\infty} p(f)df. \qquad (5.3)$$

For a car-following event, a higher average power value means the vehicle state might change fiercely and violently. Calculate the average power of the inverse of the TH and TTC so that both two values are positively correlated with the degree of aggression. In addition, jerk reflects the change rate of the acceleration or deceleration, which could be a potential behavioral characteristic of the driver [110]. Finally, Table 5.3 presents the frequency-domain features selected in this study for driving style classification.

5.2.3   PRINCIPAL COMPONENT ANALYSIS

It is difficult to get intuitive driving style classification results by analyzing multi-dimensional features directly from large naturalistic driving data, thus, the data dimensionality has to be reduced properly. Since we have constructed comprehensive indicators both from the time-domain and frequency-domain for car-following driving styles classification, then principal component analysis (PCA) is used for dimensionality reduction. The PCA approach converts a set of observations of possibly correlated variables into a set of values of linearly uncorrelated variables called principal components by using an orthogonal transformation. Assume there are a size of p feature parameters are standardized, which are expressed as a p-dimensional random vector composed as $X = (x_1, x_2, \ldots, x_p)$ and construct p new composite variables by using X, namely,

$$\begin{cases} Z_1 = l_{11}x_1 + l_{12}x_2 + \cdots + l_{1p}x_p \\ Z_2 = l_{21}x_1 + l_{22}x_2 + \cdots + l_{2p}x_p \\ \qquad \cdots \\ Z_p = l_{p1}x_1 + l_{p2}x_2 + \cdots + l_{pp}x_p. \end{cases} \qquad (5.4)$$

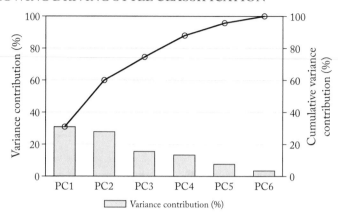

Figure 5.2: The PCA feature extraction.

Then the expression of any principal component Z_i can be written by

$$Z_i = l_{i1}x_1 + l_{i2}x_2 + \cdots + l_{ip}x_p, \ i = 1, 2, \ldots, p, \tag{5.5}$$

where Z_i denotes the new composite indicator after dimensionality reduction of original variables, $l_{ij}(j = 1, 2, \ldots, p)$ denotes the weight of the normalized features on the principal component Z_i. In the principal component analysis, the variance contribution of the i-th principal component represents how much information of the original data it contains. Generally, the first principal component is the component with the largest variance contribution.

Figure 5.2 and Table 5.3 illustrate the PCA results of the driving style classification based on the time-domain and frequency-domain features. Since the first principal component with the largest variance contribution after PCA dimensionality reduction could best synthesize the original data information, which is a fine choice for comprehensive assessment, thus, we will take the first principal component to characterize driving style in this study.

As the value of each index is proportional to the aggressiveness degree of the car-following driving style. Figure 5.2 shows that the variance contribution of the PC1 (principal component 1) is 31.11% and that of the PC2 (principal component 2) is 28.55%, both of which are significantly larger than the rest principal components. Table 5.4 further assist us to select the best principal component to build comprehensive indicators for driving style classification. Because all index values are positively correlated with the car-following aggressiveness within the PC1, while loadings of the PC2 cannot be fully consistent. Thus, the PC1 is chosen to characterize the driving style. Finally, the loadings of each index in the PC1 were are standardized before calculating the score for assessing driving style. As shown in Equations (5.6) and (5.7), the score consists of two parts, namely, the score of time-domain features S_T and the score of

Table 5.4: Principal components load matrix

Indexes	PC1	PC2	PC3	PC4	PC5	PC6
THP_s	0.433	0.207	0.770	−0.407	0.105	−0.008
$\text{TTC}_{inv}^{0.95}$	0.313	0.679	−0.321	−0.411	−0.408	−0.045
ACC_{std}	0.852	−0.370	−0.163	0.039	−0.042	0.326
THPow_{inv}	0.299	0.551	0.303	0.685	−0.214	−0.010
TTCPow_{inv}	0.211	0.792	−0.283	−0.033	0.496	−0.043
JerkPow	0.849	−0.376	−0.147	0.071	0.091	−0.322

frequency-domain features S_F, such that

$$S_T = 0.433 \times \text{THP}_s + 0.313 \times \text{TTC}_{inv}^{0.95} + 0.852 \times \text{ACC}_{std} \tag{5.6}$$
$$S_F = 0.299 \times \text{THPow}_{inv} + 0.211 \times \text{TTCPow}_{inv} + 0.849 \times \text{JerkPow}. \tag{5.7}$$

The time-domain score of one car-following event reflects its hazardous level in the time domain, while the frequency-domain score measures the overall change intensity of the states (e.g., the time-headway, TTC, jerk) during the car-following event. Obviously, both higher scores correspond to a more aggressive driving style.

5.3 CLASSIFICATION OF DRIVING STYLES BASED ON THE GAUSSIAN MIXTURE MODEL

In this chapter, we will classify the driving style scores by the GMM algorithm, then labeling each sample according to the clustering result. Finally, the effectiveness of the proposed driving style classification method will be discussed.

5.3.1 GAUSSIAN MIXTURE MODEL

Generally, clustering algorithms include the K-means algorithm, mean drift clustering methods, density-based clustering methods, the GMM, cohesive hierarchical clustering methods, and so on. Among them, the K-means and the GMM algorithm are the most commonly used unsupervised classification algorithms, however, the clustering effect of the K-means could be easily affected by the sample distribution [146]. In contrast, the GMM based on Gaussian distribution is a flexible and powerful tool for simulating complex data distributions with simple structures. The GMM obtains the posterior probability of each sample belonging to different categories, such that it further classifies the samples.

Assume that the GMM consists of K Gaussian distributions, and each Gaussian distribution corresponds to one category of the samples. The probability distribution function of the

GMM is defined by

$$p(x) = \sum_{k=1}^{K} \alpha_k N_k (x | \mu_k, \Sigma_k), \tag{5.8}$$

where

$$N_k (x | \mu_k, \Sigma_k) = \frac{1}{(2\pi)^{\frac{d}{2}} |\Sigma_k|^{\frac{1}{2}}} \exp\left(-\frac{1}{2}(x_i - \mu_k)^T \Sigma_k^{-1}(x_i - \mu_k)\right)$$

$N_k(x | \mu_k, \Sigma_k)$ is the probability density function of the k-th Gaussian distribution ($k \leq K$), μ_k and Σ_k are the mean vector and covariance matrix of $N_k(x | \mu_k, \Sigma_k)$, respectively, α_k is the mixing factor. The GMM parameters of the samples are evaluated by the Expectation–Maximization (EM) algorithm, which proceeds as follows:

(1) Input sample set and define K.

(2) Initialize the GMM parameters α_k, μ_k, Σ_k.

(3) The posterior probability of unlabeled sample x_i is generated by the k-th component is

$$p(i, k) = \alpha_k N_k (x_i | \mu_k, \Sigma_k) / \sum_{j=1}^{k} \alpha_j N_j (x_i | \mu_j, \Sigma_j). \tag{5.9}$$

(4) Update the GMM parameters using (i, k). Then calculate the maximum likelihood value of μ_k

$$N_k = \sum_{i=1}^{N} p(i, k) \tag{5.10}$$

$$\mu_k = \frac{1}{N_k} \sum_{i=1}^{N} p(i, k) x_i, \tag{5.11}$$

where N denotes the number of samples.

(5) Calculate the covariance matrix Σ_k

$$\Sigma_k = \frac{1}{N_k} \sum_{i=1}^{N} p(i, k)(x_i - \mu_k)(x_i - \mu_k)^T. \tag{5.12}$$

(6) Calculate the weights

$$\alpha_k = N_k / N. \tag{5.13}$$

(7) Repeat steps (3)–(6) until the results converged.

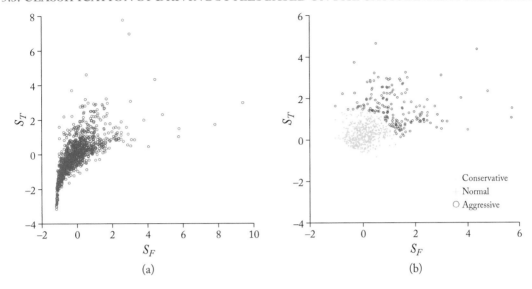

Figure 5.3: The driving style scores of samples and their clustering result.

Assuming that all scores are followed by a two-dimensional GMM with three Gaussian distributions that each distribution corresponds to one car-following driving style. Following steps (1)–(7) above to obtain the GMM parameters, the posterior probability $p(i, k)$ of x_i generated by the k-th Gaussian distributions in the GMM will be calculated. It is obvious that the exact category for the sample x_i corresponds to the maximum $p(i, k)$.

5.3.2   CLUSTERING RESULTS AND EVALUATIONS

Figure 5.3a shows the scatter plot of the original data, there are four samples whose scores are larger than 6, which are significantly higher than those of other samples. Since outliers may reduce the quality of data analysis, these four samples would be regarded as outliers and are supposed to be removed. The final clustering analysis is conducted based on the 1038 samples, and the GMM classification result is shown in Figure 5.3b.

We label the lower scores (blue) as conservative car-following driving style, the higher scores as aggressive car-following driving style, while the middle scores as normal car-following driving style. The number of samples for each category is 364 samples in the conservative, 495 samples in the normal, and 179 samples in the aggressive. Mann–Whitney rank-sum test [147] was conducted to compare the differences among indicators' mean value for each driving style's samples, and the result is shown in Table 5.5 which are all significant. That means samples with different driving styles could be well distinguished, also, the more aggressive the driving style, the larger of jerks, as well as smaller TH and TTC values.

Table 5.5: Test of the variability of indicators

Indexes	Driving Style Types	Mean Value	P-Value
Time Headway	Conservative-1	3.2348	P(1 & 2) = 0.000
	Normal-2	2.9716	P(2 & 3) = 0.000
	Aggressive-3	2.1525	P(3 & 1) = 0.000
The inverse of the TTC	Conservative-1	0.0540	P(1 & 2) = 0.000
	Normal-2	0.0638	P(2 & 3) = 0.003
	Aggressive-3	0.0762	P(3 & 1) = 0.000
Jerk	Conservative-1	0.1113	P(1 & 2) = 0.000
	Normal-2	0.2610	P(2 & 3) = 0.000
	Aggressive-3	0.2589	P(3 & 1) = 0.000

Table 5.6: Evaluation indexes and their thresholds

Indexes	Symbols	Thresholds
Time Headway	TH	2 s
The inverse of the TTC	TTC_{inv}	0.2 s^{-1}
Jerk	J	0.3 m/s^3

Next, appropriate indexes to evaluate the results of driving style clustering is necessary. Murphey et al. proposed that the jerk of a vehicle increases with the aggressiveness of the driver's driving style, and its absolute mean value is generally between 0.2–0.3 [102]. According to a study in [113], car-following is no longer safe when the time headway (TH) value is lower than 2 s, and the probability of collision could be high when the inverse of TTC is higher than 0.28. To sum up, we choose jerk, time-headway, and the inverse of the TTC as indexes to evaluate the feasibility of the driving style classification results, where the threshold values of the indexes are listed in Table 5.6. Moreover, we also calculate the proportion of the time-headway below the corresponding threshold, and the proportion of the inverse of the TTC and jerk above their thresholds in each category. Large proportions indicate that samples maintain a high-risky time headway and have low TTCs during the car-following phase, meanwhile, tend to accelerate and decelerate sharply, which corresponds to the aggressive driving style.

As shown in Figure 5.4, for the conservative car-following driving style, the proportion of the inverse of the TTC and jerk above their thresholds are 0.34% and 15.16%, respectively, while the proportion of time-headway below the threshold is 15.45%, which shows that conservative samples generally maintain a relatively safe car-following state and their acceleration and deceleration are small. Moreover, for normal car-following driving style, the proportions are twice

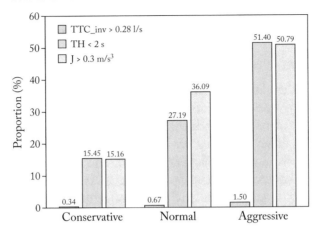

Figure 5.4: Proportion results of different driving styles.

higher than those of the conservative, while the proportions in the aggressive driving style are more than three times higher compared with the conservative. In a word, with the increase of the driving styles' aggressiveness, the proportion of high-risk time-headway, low TTCs, and large jerks would also increase, which indicates that the driving style classification method proposed in this study is satisfying.

5.4   INFLUENCE OF THE DRIVING ENVIRONMENTAL FACTORS ON CAR-FOLLOWING DRIVING STYLE

In the previous chapter, a driving style classification method based on principal component analysis (PCA) and Gaussian mixture model (GMM) clustering is proposed, and representative indicators to evaluate the car-following driving style are selected. Compared with the K-means clustering, the GMM is confirmed with better stability in driving style classification. However, there are still some limitations to use the NGSIM dataset. The driving scenario of the NGSIM dataset is collected on one segment of highway with a similar surrounding environment and the data were collected almost at the same time in a single day. The classification result of car-following driving style might be useful in this scene, but it might be unknown when the driving scenario setting such as the road types and weather were changed. Previous studies have found that the driving environments affect the driving behavior of drivers and driving styles. Lyons et al. suggested that driving patterns are connected with the urban structure and local environment [179]. Moreover, some researchers predict aggressive driving styles based on drivers' activity and environment parameters [180]. As regards modeling driving style depending on the driving environment, some researchers distinguish driving styles among urban, suburban, and highway driving environments [181]. Therefore, the classification method of the car-following driving style might be sensitive to the driving environment.

5.4.1   EXTRACTION OF CAR-FOLLOWING EVENTS

In general, to further discuss the influence of the driving environment on car-following driving style, car-following events are extracted from a self-acquisition driving dataset with different driving environment settings. Such that we can analyze the differences of car-following driving styles among different driving environments, which is based on the previously proposed car-following driving style classification method. The involved self-acquisition driving dataset in this chapter is collected by an economic-cost driving simulator based on the PreScan® and the Logitech G29. Regarding the driving environment, road types and weather conditions are considered separate scenes. To be specific, the road type is divided into highway and city roads, while the driving weather condition is divided into sunny weather and foggy weather. The driving simulator records the speed and acceleration of the vehicle in all directions while driving. And the relative distance and the relative speed to the preceding car are also recorded. Finally, a total of 693 car-following events from 10 volunteer participants are obtained, in which the extraction criteria are listed below.

(1) The relative distance to the preceding car is less than 100 m in the highway scenario, while less than 50 m in the city scenario.

(2) The duration of the car-following event lasted more than 10 s.

(3) The preceding car is not lane changing in the car-following situation.

(4) The self-car is not lane changing in the car-following situation.

5.4.2   CAR-FOLLOWING DRIVING STYLE CLUSTERING REGARDLESS OF THE DRIVING ENVIRONMENT

Before considering the influence of the driving environment, we involve all car-following events at first and verify whether the driving style classification method proposed in the previous chapter is suitable for the current self-acquisition dataset. The time-domain and frequency-domain features of the TH (time headway), TTC (time to collision), and jerk are adopted for the PCA. Multivariate characteristic parameters are dimensionally reduced by the PCA, and the information of the first principal component with the largest variance contribution rate is used. Then, the score of the time-domain features and the frequency-domain features are calculated by the PCA loading matrix.

The GMM is adopted directly as the method of car-following driving style classification method. After removing 2 outlier samples, the rest of 691 car-following events are reminded for the following analysis. The clustering results are shown in Figure 5.5, in which the car-following events with low scores are classified as conservative, while the car-following events

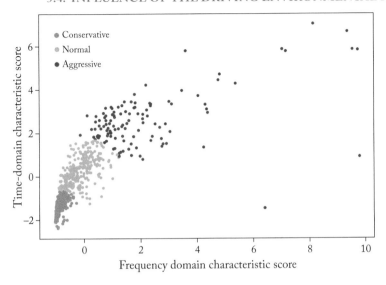

Figure 5.5: Clustering results of car-following driving style in self-acquisition datasets.

Table 5.7: Mann–Whitney rank-sum test of driving style of car-following events

Indicators	Driving Style	Mean Value	P-Value
Time headway	Conservative-1	2.939	P(1 & 2) = 0.000
	Normal-2	2.038	P(2 & 3) = 0.000
	Aggresive-3	1.380	P(3 & 1) = 0.000
Time to collision	Conservative-1	19.186	P(1 & 2) = 0.000
	Normal-2	9.605	P(2 & 3) = 0.000
	Aggresive-3	4.990	P(3 & 1) = 0.000
Jerk	Conservative-1	0.175	P(1 & 2) = 0.000
	Normal-2	0.643	P(2 & 3) = 0.000
	Aggresive-3	1.392	P(3 & 1) = 0.000

with high scores as aggressive driving style, and the rest car-following events with moderate scores as normal driving styles. Finally, the number of car-following event samples of various driving styles is 291 conservative samples, 281 moderate samples, and 119 aggressive samples, respectively. We conduct Mann-Whitney rank-sum test for the mean difference values among indicators within each driving style, the results are shown in Table 5.7, which shows that there are significant differences among the TH, TTC, and jerk for each driving style, indicating the car-following driving styles are well distinguished through the GMM clustering.

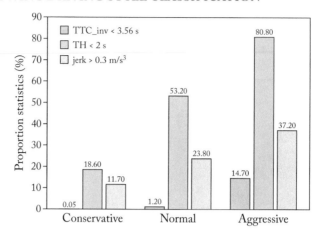

Figure 5.6: Proportion results of the indicators for each car-following driving style.

Moreover, the proportion of the low TH and TTC, as well as the large jerk are adopted to verify the feasibility of the driving style classification, the threshold of low TH is 2 s and TTC is 3.56 s, respectively, and that for large jerk is 0.3 m/s^3. The results are shown in Figure 5.6, where the proportion of low TTC is 0.05% for the conservative samples, while the corresponding proportion of low TH and large jerk is 18.6% and 11.7%, respectively. It is clear that the overall jerk range of conservative samples is small, meanwhile, safe following states are maintained. In contrast, the proportion of each index for the aggressive driving style is the largest. Thus, the driving style classification method developed in the previous chapter not only performs well on the NGSIM dataset but is also valid on our self-acquisition driving dataset.

5.4.3   CAR-FOLLOWING DRIVING STYLE CLUSTERING CONSIDERING THE DRIVING ENVIRONMENT

A total of 691 car-following events with outliers removed are divided by driving environments, based on the road type, and these car-following events are divided into highway (411 samples) and city (280 samples). Moreover, they could be also divided into sunny weather (330 samples) and foggy weather (361 samples) according to the weather conditions. The car-following driving style clustering is conducted by separating car-following events from the driving environment, and the influence of different road types and weather conditions on car-following driving styles are analyzed.

The clustering results considering the driving environment are shown in Figure 5.7 and Table 5.8, and the indicators' mean value for each car-following driving style is presented in Table 5.9. Moreover, the proportion statistics of the low TH and TTC, as well as large jerks for different driving styles while considering the driving environment are shown in Table 5.10.

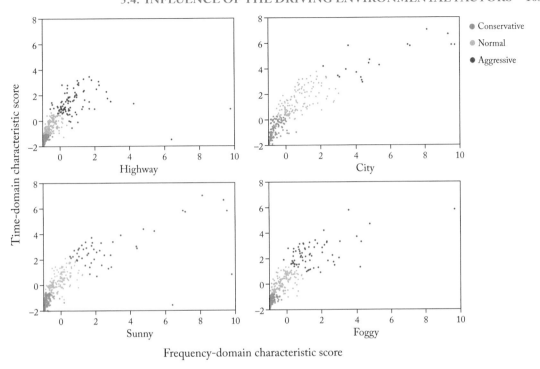

Figure 5.7: The driving style clustering results considering the driving environment.

Table 5.8: The proportion statistics of driving styles considering the driving environment

Environment Factors		Conservative	Normal	Aggressive	Silhouette Coefficient
Road types	Highway	171	146	94	0.481
	City	136	126	18	0.520
Weather	Sunny	151	131	48	0.505
	Foggy	152	139	70	0.525

The results show that indicators of the car-following event with the same style may differ greatly in different environments. More specifically, the influence of road types on driving style is as follows:

(1) In the scenario of the highway, the means TH of all driving styles are larger than those of the city, as well as the proportions of low TH. This may be due to the heavy traffic on city roads, which can be verified in the self-acquisition dataset. It is shown that the average

Table 5.9: Driving style difference test considering the driving environment

Indicators	Driving Style	Road Types			Weather		
		Highway	City	P-value	Sunny	Foggy	P-value
Mean time	Conservative-1	3.190	2.324	P(1&2)=0.000	2.678	3.138	P(1&2)=0.000
headway(s)	Normal-2	2.409	1.468	P(1&2)=0.000	1.912	2.144	P(1&2)=0.000
	Aggressive-3	1.883	0.947	P(1&2)=0.000	1.425	1.343	P(1&2)=0.000
Mean	Conservative-1	22.390	14.234	P(1&2)=0.000	17.887	20.351	P(1&2)=0.000
TTC (s)	Normal-2	11.663	6.527	P(1&2)=0.000	8.622	10.388	P(1&2)=0.000
	Aggressive-3	7.355	2.779	P(1&2)=0.000	4.498	5.337	P(1&2)=0.000
Mean	Conservative-1	0.161	0.312	P(1&2)=0.000	0.201	0.164	P(1&2)=0.000
Jerk (m/s^2)	Normal-2	0.433	1.033	P(1&2)=0.000	0.680	0.640	P(1&2)=0.000
	Aggressive-3	1.087	1.759	P(1&2)=0.000	1.486	1.321	P(1&2)=0.000

Table 5.10: Proportion statistics of indicators exceeding the threshold while considering the driving environment

Indicators	Driving Style	Road Types		Weather Conditions	
		Highway	City	Sunny	Foggy
TH < 2 s (%)	Conservative	0.105	0.404	0.233	0.157
	Normal	0.390	0.778	0.591	0.502
	Aggressive	0.614	0.926	0.795	0.827
Jerk > 0.3 m/s^2 (%)	Conservative	0.117	0.142	0.127	0.111
	Normal	0.197	0.318	0.249	0.245
	Aggressive	0.337	0.397	0.389	0.368
TTC < 3.56 s (%)	Conservative	0.000	0.002	0.000	0.000
	Normal	0.006	0.074	0.020	0.010
	Aggressive	0.061	0.400	0.227	0.121

distance of car-following events on the highway is 67.8 m, while that on a city road is only 33.0 m ($P < 0.05$).

(2) The average jerk of car-following events on the highway is smaller than that in cities. This may be caused by the more complex driving scene in the city, which may result in a more sharp brake situation and greater jerk. In the self-acquisition dataset, the average longitudinal jerk of highway of car-following events is 0.33 m/s^3, while that of city roads is 0.37 m/s^3, but no significant difference ($P = 0.64$) is reported. Meanwhile, when analyzing

the proportion of high jerks, the proportion for each driving style on highway roads is less than that on city roads.

(3) In the highway scenario, the mean TTCs of the car-following events in all driving styles are smaller than the city roads, and the proportion of low TTC on the highway is larger than the city. As mentioned above, this may be caused by the greater density of vehicles and smarter car-following distance on city roads. In addition, the city scene is more complex, which may cause more situations of sudden braking, and result in smaller TTC.

The influence of weather on car-following driving style is listed as follows.

(1) On sunny weather, the means TH of conservative and normal driving styles are smaller than those of foggy weather. The proportion of low TH in the sunny weather is larger than that in the foggy one of conservative and normal driving styles. The reason could be that drivers often keep a closer car-following distance and higher speed on sunny weather because of better visibility. It is also proved in the self-acquisition dataset that the average distance of car-following events in the sunny scenario is 52 m, while that in the foggy scenario is 62 m ($P < 0.05$). However, when focusing on aggressive style events, the results are opposite for means TH and proportion of low TH.

(2) The means TTC of all driving styles is lower on sunny weather than foggy weather. Besides, in the sunny scenario, the proportion of the TTC below the threshold is larger than that on foggy days. As mentioned above, this could be due to drivers' tendency to keep a more aggressive driving state and smaller car-following distance in sunny scenarios because of higher visibility, which is similar to the analysis of the TH.

(3) The mean jerks of all driving styles are greater in sunny weather than those in foggy weather, and the proportions of large jerks are also greater in the sunny scenario. Therefore, it can be seen that drivers are less inclined to accelerate and brake sharply in foggy weather, which is reasonable to be safe driving. However, the difference in the proportion of large jerks between sunny and foggy weather is not very significant.

5.5 SUMMARY

Overall, a driving style classification method based on the PCA and the GMM is proposed and validated. Based on the NGSIM naturalistic driving databased, selective features both from the time-domain and frequency-domain are obtained for the car-following driving style. After reducing the dimensionality of the features through the PCA, the comprehensive evaluation mechanism for car-following driving style is established based on the first principal component. Moreover, the GMM is applied to classify the driving styles of each sample. Based on the statistical proportions of the low TH and the TTC, as well as large jerks, it shows that a reasonable classification result for car-following driving styles is obtained.

In addition, this study further conducts analysis based on a self-acquisition simulative driving dataset while considering the driving environment. On the one hand, the validity of the driving style classification method proposed in this study can be further verified. On the other hand, the influence of the driving environment on car-following driving style is analyzed. The study has proved that the road type and the weather condition have a great impact on the car-following driving style. In general, there is a significant difference in driving styles among different driving environments. For every driving condition and style, city driving has smaller THs and TTCs but larger jerks than highway driving, which may be related to the more complex driving conditions in the city with smaller gaps among cars. Meanwhile, for most of the conditions, driving in sunny weather conditions has smaller THs and TTCs but larger jerks than foggy weather conditions, which may be related to the better visibility in clear conditions. Therefore, sunny weather and city road have a positive impact on aggressive driving styles, and then drivers who driving in clear weather conditions and city road environment tend to perform more aggressive driving styles.

CHAPTER 6

Driving Behavior Analysis Based on Naturalistic Driving Data

Since the driver's risky driving behaviors cause the majority of traffic accidents, and the driver's demographic characteristics, personality traits, and psychological factors could affect risky driving behaviors. Therefore, it is very helpful to explore the influence mechanism of driver's demographic characteristics and psychological factors on risky driving behaviors for traffic safety. The main purpose of this chapter is to take the driving experience as a continuous variable and explore its relationships among sensation seeking, risk perception, as well as risky driving behaviors. Incorporated with large-scale questionnaire surveys in the Strategic Highway Research Program 2 (SHRP 2), a meditation model of driver's characteristics on risky driving behaviors is constructed via the Structural Equation Model (SEM) to verify the relationship. After taking gender as an interactive variable, a moderated mediation model is constructed to explore the differences in driving experience on risky driving behaviors between different genders, which is also depicted in Figure 6.1.

Furthermore, we build a classification model of driver's driving risk by using the Random Forest (RF) method, in which the Crash and Near-crash (CNC) rate is adopted as a training label and clustered into three risk levels including the low-risk, the middle-risk, and the high-risk, by the K-means clustering method. While drivers' demographics (age, cumulative driving years, gender), sensation seeking, risk perception, and risky driving behaviors are the input variables of the classification model. Thus, the driving risk level of drivers is expected to be identified in the RF classifier. A quick flow-chart of this study is depicted in Figure 6.2, and the main contributions are summarized in brief as follows:

1. The influences of driving experience on risky driving behaviors mediated by sensation seeking and risk perception are investigated.

2. The influence trends of driving experience on risky driving behaviors between different genders are further explored and analyzed.

3. A classification model of the driver's driving risk is proposed based on the self-reported questionnaires of the driver's demographics, sensation seeking, risk perception, and risky driving behaviors by using the RF algorithm.

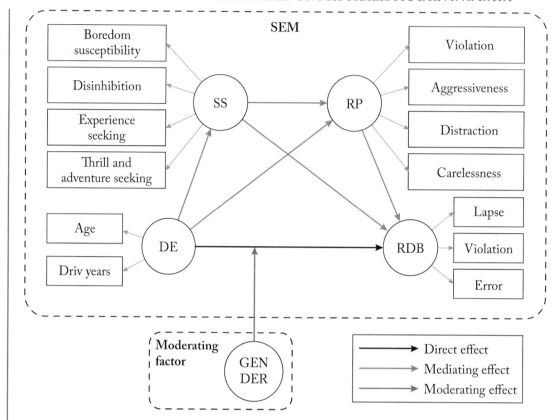

Figure 6.1: The hypothesized SEM including mediating effects and moderating effects. *Note:* Blackline represents the direct effect; Orange line represents the mediating effect; the blue line represents the moderating effect; DE: driving experience, SS: sensation seeking, RP: risk perception, RDB: risky driving behaviors; moderating factor: gender.

6.1   SUBJECTIVE SELF-REPORTED RISKY DRIVING BEHAVIORS ANALYSIS

In this section, we will concentrate on the influence of driver's demographical factors (driving experience, gender) on risky driving behaviors, and explore the role of driver's psychological factors (sensation seeking, risk perception) play in this influence.

6.1.1   DATA ACQUISITION

Data from the Strategic Highway Research Program 2 (SHRP2) are used in this chapter. The SHRP 2 Naturalistic Driving Study, which is one of the largest naturalistic driving-behavior studies to date, monitored over 3,500 participants from 2010–2013 in the United States. One

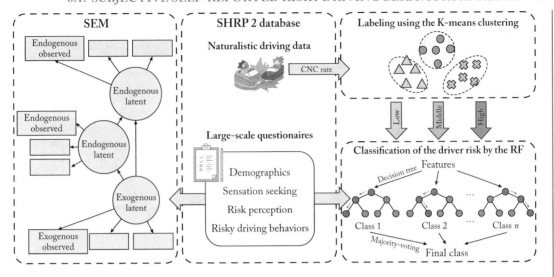

Figure 6.2: The brief flow of the SEM and the Random Forest classification model in this study. *Note.* CNC: crash and near-crash.

way to provide the public with access to SHRP 2 data is the SHRP 2 Insight website (https: //insight.shrp2nds.us/), which allows researchers to browse de-identified driving data and build queries to search what they are interested in.

The data in this chapter were extracted from self-reported questionnaires in the driver branch and naturalistic driving data in the event branch of the SHRP 2. Before collecting the naturalistic driving data, drivers were asked to fill out a series of questionnaires, such as the drivers' demographical questionnaire, sensation-seeking scale (SSS), risk perception questionnaire (RPQ), and driving behavior questionnaire (DBQ). The naturalistic driving data includes the crash, near-crash, and baseline records. Besides, the CNC records are extracted by analyzing the naturalistic driving data. The CNC rates in this study are calculated by using the CNC records and baseline records, which are necessary for comparisons with crashes and near-crashes. These baselines in the SHRP 2 were randomly selected with a goal of 20,000 baselines and a minimum of 1-baseline per driver. The number of baselines per driver is proportional to the total driving time larger than 8.064 km/h (5-mph). Therefore, the driver's CNC rate could be represented by the number of crash and near-crash records per baseline records.

In summary, the variables involved in this chapter are driver's demographics, including driving experience (represented by age and cumulative driving years) and gender; sensation seeking and risk perception; risky driving behaviors; and the CNC rate.

Participants involved in this chapter comprise approximately half female (1,701) and half male (1,499) drivers, ages range from 16–99, and cumulative driving years range from 0–76. After deleting the drivers whose deletion of items in questionnaires are more than 20%, the

Table 6.1: Statistical table of driver demographics from the SHRP 2 Insight

Variables	Category	Frequency	Percent	Variables	Category	Frequency	Percent
	16–19	529	0.168		0–4	776	0.246
	20–24	721	0.229		5–9	603	0.191
	25–34	405	0.129		10–19	339	0.108
Age	35–49	351	0.111	Cumulative driving years	20–34	338	0.107
	49–64	389	0.123		34–49	393	0.125
	65–79	553	0.176		49–64	551	0.175
	79–	188	0.060		64–	136	0.043

samples of the remaining 3,150 drivers are included in the analysis, where the number of female drivers is 1499 which accounts for 47.5% of the total number, while the number of male drivers is 1651 which accounts for 52.4% of the total number. On account of the fact that most American drivers are used to getting their licenses at age 16, there could be a strong correlation between age and cumulative driving years. Indeed, age is positively and significantly related to the cumulative driving years ($r = 0.94$, $p < 0.01$) according to the SHRP 2 data, which indicates they are related but not entirely overlapping. Therefore, similar to the study of Constantinou et al. [116], age and cumulative driving years are regarded as two observed variables for representing the latent variable "driving experience" in the chapter. Besides, it should be noted that the age and cumulative driving years would be taken as continuous variables in the study according to the original SHRP 2 Insight data. Nevertheless, for the convenience of presenting a descriptive statistic, the statistical summary of drivers' demographic factors including the age groups, and cumulative driving years, is shown in Table 6.1.

Besides, considering small samples will easily lead to convergence failure, inappropriate solutions (illegal estimates), low parameter estimates, and incorrect standard errors, Bentler and Chou suggest that the sample size should be at least 5 times the estimated parameter when using the SEM (in the case of normal, no omission value and no extreme value, otherwise the sample size should be at least 15 times) [148]. Kline et al. also recommend that the number of samples should up to 20 times [149]. As incorporating with the large-scale data of the SHRP 2 Insight, the sample size of this study meets those requirements.

6.1.2   MEASUREMENTS OF RISKY DRIVING BEHAVIORS BASED ON THE DBQ

To study the influence of driver's subjective factors on risky driving behaviors, in addition to looking for driver's characteristics that have an impact on driving behaviors, it is also necessary to specify a set of measurement standards for risky driving behavior, to conduct quantitative analysis on driver's risky driving behaviors. Therefore, it is urgent and necessary to develop a methodology that can measure the frequency of these behaviors they committed, and determine which actions are the most likely to predict traffic collision involvement [149].

The Driver Behavior Questionnaire (DBQ) to measure the driver's driving behaviors were used in this chapter. The DBQ developed by the Manchester driver behavior group is an often-used toolbox to measure abnormal driving behaviors [128]. So far, the self-reported Manchester DBQ is one of the most widely used approaches for measuring behavior and bad habits in daily driving [150]. Especially, the three-component structure of the DBQ, which includes *Harmless lapse*, *Dangerous error*, and *Violation*, is widely used in earlier studies. Violation is defined as deliberate deviations from those practices believed necessary to maintain the safe operation of a potentially hazardous system. Lapse and error are defined as the failure of planned actions to achieve their intended consequences. Lapse is the unwitting deviation of action from intention, while error is the departure of planned actions from some satisfactory path toward a desired goal [151]. Moreover, subsequent researchers developed a four-component structure. They have subdivided the *Violation* into *Ordinary violation* and *Aggressive violation* that has emotional components [152].

It is found that the DBQ score was correlated with objectively measured risky driving behaviors to certain extents [153]. Besides, previous studies have shown that the DBQ score is correlated with driving crashes. Furthermore, the combination of error scores could predict crashes [150]. There also seems to be a strong correlation between age and risky driving behaviors, and young drivers are more likely to be engaged in risky driving behaviors in all categories of the DBQ [154].

6.1.3   FACTOR ANALYSIS

Due to the differences in policies and cultures, the structures of drivers' questionnaires will vary in different regions. For example, Lajunen et al. analyzed the DBQ in Finland and the Netherlands through the four-factor structures (*Aggressive violation, Ordinary violation, Error,* and *Lapse*) [155]. However, the results of the four-factor structures show that the structure of the DBQ found in Finland and the Netherlands was congruent but not perfect with the target structure found in Britain, which indicates that the structure of the DBQ varies from different areas and even different times. Therefore, the structure should be analyzed and carefully chosen according to practical situations to ensure the validity of the scale when using the DBQ. Specifically, we conducted exploratory factor analysis (EFA) first to reduce the dimensions of DBQ and RPQ scales, to obtain several sub-dimensions that can best represent the questionnaire. Moreover, it is also necessary to conduct EFA before establishing the SEM [156].

To eliminate the collinearity of items in questionnaires, principal component analysis (PCA) is used to conduct exploratory factor analysis on 24 items in the DBQ and 32 items in the RPQ. Considering the SSS in SHRP 2 has been processed in advance and is divided into four dimensions [120], namely *Boredom, Inhibition, Experience,* and *Thrill,* thus it will not be included in the EFA. However, the confirmatory factor analysis (CFA) of the SSS will be still conducted in subsequent studies. The dimensionality reduction results are shown in Tables 6.2a, 6.2b, and 6.2c, and item scores greater than 0.30 are retained.

In the DBQ analysis, the suitability of the data for posterior analysis uses the Kaiser Meyer–Olkin test (.898) and Bartlett's sphericity test ($\chi^2(66) = 11814.799, P < 0.001$). Due to the universal use of the three-component structure and four-component structure of DBQ, we compare the factor analysis results of these two structures. It can be found that the DBQ divided into four dimensions explains 38.18% variance variation. As shown in Tables 6.2a, 6.2b, and 6.2c the four dimensions of the DBQ include the frequency of taking dangerous errors (*Error*), violating the traffic rules (*Violation*), committing harmless errors (*Lapse*), and engaging in aggressive driving behaviors (*Aggressiveness*). Meanwhile, the DBQ divided into three dimensions explains 33.69% variance variation. As shown in Table 6.3, the three dimensions of the DBQ include the frequency of taking dangerous errors (*Error*), violating the traffic rules (*Violation*), and committing harmless errors (*Lapse*). The structure of the DBQ will be decided in the CFA subsequently.

While for the analysis of the RPQ, the suitability of the data for posterior analysis uses the Kaiser Meyer–Olkin test (.977) and Bartlett's sphericity test ($\chi^2(66) = 74491.591, P < 0.001$), as shown in Table 6.4. The RPQ is divided into three dimensions. Moreover, since the fourth dimension with an eigenvalue of 0.968 approaches 1, it is also included in the analysis as well. The extracted four dimensions explained a total of 64.38% variance variation. The four dimensions of RPQ include perception of aggressive driving behaviors (*Aggressiveness*, e.g., racing), common violation (*Violation*, e.g., breaking red-light), engaging in secondary driving tasks while driving (*Distraction*, e.g., talking with a near passenger), and committing careless driving habits (*Careless*, e.g., fail to check tires before driving). These dimensions will be incorporated into the model as observed variables of the SEM.

According to the dimensionality reduction, these three questionnaires of the DBQ, SSS, and RPQ are divided into several sub-dimensions, respectively, which are used as observed variables of potential variables. The SSS is divided into four sub-dimensions, namely, the *BS* (boredom susceptibility, a dislike of repetition of experience), *DIS* (disinhibition, the loss of social inhibitions), *ES* (experience-seeking, its essence is "experience for its own sake."), and the *TAS* (thrill and adventure seeking, a desire to engage in outdoor sports or other activities.). The RPQ is divided into four sub-dimensions, including the *Violation* (the perception of taking deliberate illegal traffic rules of behavior), *Aggressiveness* (the perception of engaging in aggressive driving behavior), *Distraction* (the perception of taking the second driving task in the process of driving), and *Carelessness* (the perception of careless driving habits). Since the three-component and four-component structures are commonly used in the DBQ study, they will be verified separately by the CFA to select the most suitable structure for current samples.

Subsequently, questionnaires after the dimensionality reduction are verified by the confirmatory factor analysis (CFA) to test the reliability and validity of the sub-dimension. The

Table 6.2a: Rotated component matrices of the four-component DBQ *(Continues.)*

No	Items: How often the participant has …	Error	Violation	Lapse	Aggressiveness
21	missed yield signs and narrowly avoided colliding with traffic that had the right of way (*E*)	.643			
9	failed to notice that pedestrians are crossing when turning onto a side street from the main road (*E*)	.535			
24	braked too quickly on a slippery road or steered the wrong way into a skid (*E*)	.534			
13	when turning right, nearly hit a bicyclist who was riding right alongside them (*E*)	.514			.309
17	underestimated the speed of an oncoming vehicle when attempting to pass a vehicle in their lane (*E*)	.499			
22	failed to check their rearview mirror before pulling out, changing lanes, etc. (*E*)	.493	.325		
18	hit something when backing up that they had not previously seen (*E*)	.490			
11	misread the signs and turned the wrong direction on a one-way street (*E*)	.488			
14	attempted to turn onto the main road and been paying such close attention to traffic entering that they nearly hit the car in front of them that is also waiting to turn (*E*)	.449			
20	gotten into the wrong lane approaching an intersection (*L*)	.396		.375	
2	become impatient with a slow driver in the fast lane and passed on the right (*V*)		.719		
3	driven especially close to a car in front as a signal to the driver to go faster or get out of the way (*V*)		.704		
12	disregarded the speed limits late at night or early in the morning (*V*)		.624		

Table 6.2a: (*Continued.*) Rotated component matrices of the four-component DBQ

No	Items: How often the participant has ...	Error	Violation	Lapse	Aggressiveness
4	attempted to pass someone that they hadn't noticed was making a left turn (*V*)		.489		
8	crossed an intersection knowing the traffic lights have already changed from yellow to red (*V*)		.475		
15	driven even though they might be over the legal blood alcohol limit (*V*)		.403		
5	forgotten where they left their car in a parking lot (*L*)			.709	
7	realized they had no clear recollection of the road along which they have been traveling (*L*)			.670	
19	found themselves intending to drive to destination A, but find themselves on the road to destination B perhaps because destination B is a more common destination (*L*)			.606	
6	turned on one thing, such as headlights, when they meant to switch on something else such as the windshield wipers (*L*)			.548	
1	attempted to drive away from traffic lights in the wrong gear (*L*)				
10	been angered by another driver's behavior and caught up to them intending to give them a piece of their mind (*A*)				.736
16	had an aversion to a particular class of road users and indicated their hostility by whatever means (*A*)				.705
23	got involved in unofficial "races" with other drivers (*A*)		.300		.493

Note: Rotated coefficients less than 0.3 have been hidden; *E* = Error, *V* = Violation, *L* = Lapse, *A* = Aggressiveness.

Table 6.2b: Rotated component matrices of the three-component DBQ (*Continues.*)

No	Items: How often the participant has …	Components		
		Error	Violation	Lapse
21	missed yield signs and narrowly avoided colliding with traffic that had the right of way (*E*)	.639		
13	when turning right, nearly hit a bicyclist who was riding right alongside them (*E*)	.581		
11	misread the signs and turned the wrong direction on a one-way street (*E*)	.512		
9	failed to notice that pedestrians are crossing when turning onto a side street from the main road (*E*)	.510		
17	underestimated the speed of an oncoming vehicle when attempting to pass a vehicle in their lane (*E*)	.484		
14	attempted to turn onto the main road and been paying such close attention to traffic entering that they nearly hit the car in front of them that is also waiting to turn (*E*)	.471		
24	braked too quickly on a slippery road or steered the wrong way into a skid (*E*)	.465		
18	hit something when backing up that they had not previously seen (*E*)	.463		
22	failed to check their rearview mirror before pulling out, changing lanes, etc. (*E*)	.394		
3	driven especially close to a car in front as a signal to the driver to go faster or get out of the way (*V*)		.725	
2	become impatient with a slow driver in the fast lane and passed on the right (*V*)		.672	
12	disregarded the speed limits late at night or early in the morning (*V*)		.584	
10	been angered by another driver's behavior and caught up to them intending to give them a piece of their mind (*V*)		.531	
23	got involved in unofficial "races" with other drivers (*V*)		.500	
4	attempted to pass someone that they hadn't noticed was making a left turn (*V*)		.498	

Table 6.2b: (*Continued.*) Rotated component matrices of the four-component DBQ

No	Items: How often the participant has …	Components		
		Error	Violation	Lapse
16	had an aversion to a particular class of road users and indicated their hostility by whatever means (*V*)	.376	.413	
8	crossed an intersection knowing the traffic lights have already changed from yellow to red (*V*)		.409	.346
15	driven even though they might be over the legal blood alcohol limit (*V*)		.377	
5	forgotten where they left their car in a parking lot (*L*)			.665
7	realized they had no clear recollection of the road along which they have been traveling (*L*)			.630
19	found themselves intending to drive to destination A, but find themselves on the road to destination B perhaps because destination B is a more common destination (*L*)			.583
6	turned on one thing, such as headlights, when they meant to switch on something else such as the windshield wipers (*L*)	.335		.476
20	gotten into the wrong lane approaching an intersection (*L*)	.353		.427
1	attempted to drive away from traffic lights in the wrong gear (*L*)			

Note: Rotated coefficients less than 0.3 have been hidden; E = Error, V = Violation, L = Lapse.

Table 6.2c: Rotated component matrices of the four-component DBQ (*Continues.*)

No	Items: The participants associated risk with …	Components Aggressiveness	Violation	Distraction	Careless
17	drinking alcohol or using recreational drugs while driving (*A*)	.835			
1	running a red light (*A*)	.763			
16	driving shortly after drinking alcohol or using recreational drugs (*A*)	.738			
3	taking risks while driving because it's fun, like driving fast on curves or "getting air" (*A*)	.698			
2	driving when sleepy and finding it hard to keep their eyes open (*A*)	.653			
23	racing other cars or drivers (*A*)	.646			
31	passing where visibility is obscured (*A*)	.646			
7	not yielding the right of way (*A*)	.611	.474		
26	driving more than 20 mph over the limit (*A*)	.605			
24	not checking their rearview mirror when passing another car or merging onto the highway (*A*)	.579			
27	no yielding to pedestrians (*A*)	.560			.425
5	going through a stop sign without stopping (*V*)	.538	.518		
14	trying to be the first off the line when a light turns green (*V*)		.766		
15	accelerating when a traffic light turns yellow (*V*)		.698		
4	changing lanes suddenly to get ahead in traffic (*V*)	.432	.649		
8	making illegal turns (*V*)		.645		
32	not making a full stop at a stop sign (*V*)		.639		

Table 6.2c: (*Continued.*) Rotated component matrix of the four-component DBQ

No	Items: The participants associated risk with …	Components			
		Aggressiveness	Violation	Distraction	Careless
11	following emergency vehicles when the siren is on (*V*)		.621		
13	passing other cars on the right side or the shoulder of the road (*V*)		.611		
18	cutting off, honking, or yelling at other drivers who drive too slowly or cut you off (*V*)		.606		
6	speeding for the thrill of it (*A*)	.542	.590		
19	driving to reduce tension (*V*)		.578	.408	
11	taking more risks because you are in a hurry (*A*)	.445	.558		
29	turning without signaling (*L*)		.542		.451
10	following a car very closely or tailgating (*A*)	.483	.533		
25	driving 10-20 mph over the limit (*V*)		.524		
12	driving at their normal speed during bad driving conditions such as road construction, rain, snow, or ice (*V*)		.432		
21	taking your eyes off the road to adjust the CD player or pick something up from the floor (*D*)			.803	
20	doing other things while driving, like using a cell phone, eating or drinking, putting on makeup, reading things, or smoking cigarettes (*D*)			.784	
22	taking your eyes off the road to talk to passengers (*D*)			.742	
28	driving without wearing a safety belt (*L*)				.777
30	driving with badly worn tires (*L*)	.434			.459

Note: Rotated coefficients less than 0.3 have been hidden; *A* = Aggressive, *V* = violation, *D* = Distraction, *C* = Careless.

Table 6.3: Confirmatory factor analysis, including composite reliability and convergent validity

Latent Variables	Observed Variables	Parameters of Significant Test				STD. Loading	Item Reliability	Composite Reliability	Convergent Validity (AVE)
		Estimate	S.E.	EST/S.E.	p-value				
SS	BY								
	BS	1				.534	.285	.704	.378
	DIS	1.922	.086	22.451	***	.76	.578		
	ES	1.253	.058	21.436	***	.586	.343		
	TAS	1.777	.086	20.749	***	.552	.305		
RP	BY								
	Violation	1				.959	.920	.906	.708
	Aggressiveness	.791	.010	78.339	***	.880	.774		
	Distraction	.887	.016	55.172	***	.743	.552		
	Carelessness	.949	.016	58.248	***	.765	.585		
RDB (three-component)	BY								
	Error	1				.811	.658	.716	.463
	Violation	0.893	.052	17.102	***	.550	.303		
	Lapse	1.127	.063	17.992	***	.654	.428		
RDB (four-component)	BY								
	Error	1				.752	.566	.705	.381
	Violation	1.354	.056	24.350	***	.593	.352		
	Aggressiveness	.690	.034	20.285	***	.458	.210		
	Lapse	1.293	.052	24.976	***	.630	.397		

Note: ***: p-value < 0.001; S.E.: standard error; STD: standardized; EST: estimate; DE: driving experience; SS: sensation seeking; RP: risk perception; RDB: risky driving behaviors; BS: boredom susceptibility; DIS: disinhibition; ES: experience seeking; TAS: thrill and adventure-seeking. The factors of the RDB with the four-component structure indicate that it is abandoned in the following study.

Table 6.4: Analysis of discriminant validity

Variables	Convergent Validity	Discriminant Validity			
	AVE	DE	SS	RP	RDB
DE	.962	**.981**			
SS	.378	−.495	**.615**		
RP	.708	.438	−.372	**.841**	
RDB	.463	−.184	.326	−.251	**.680**

Note: The bold diagonal is the square root of Average variance extraction (AVE), and the lower triangle is the Pearson correlation; DE: driving experience; SS: sensation seeking; RP: risk perception; RDB: risky driving behaviors.

CFA is conducted to test whether the risky driving behaviors (RDB), risk perception (RP), and sensation seeking (SS) could explain their observed variables effectively, such that to correct observed variables that do not belong to these latent factors. More specifically, the reliability and validity of three questionnaires were analyzed, which can evaluate the explanatory power of questionnaires on items.

The performance indexes can be found in Table 6.3, as we can observe, standardized factor loadings in the SS with four observed variables range from 0.534 (*BS*) to 0.76 (*DIS*), standardized loadings in the RP range from 0.743 (*Distraction*) to 0.959 (*Violation*), while loadings in the RDB with four-component structure range from 0.458 (*Aggressiveness*) to 0.752 (*Error*). Since the factor loading of *Aggressiveness* in the RDB with a four-component structure is only 0.458, which indicates that its explanatory ability with a four-structure structure is unsatisfactory. Thus, the three-component structure of the RDB is adopted in the following analysis, whose loadings range from 0.550 (*Violation*) to 0.811 (*Error*),

Among them, the minimum factor loading is the *Violation* in the RDB (.550), while the maximum one is the *Violation* in the RP (.959). As most of the factor loadings are larger than 0.6, it indicates that these observed variables can be explained by the latent variables well. However, a few factor loadings range between 0.5–0.6, which shows that their explanatory ability is not perfect but still acceptable.

The composite reliabilities of the three questionnaires are 0.704 (SS), 0.906 (RP), and 0.716 (RDB), respectively, which shows that the consistency of observed variables in the three questionnaires is ideal. The convergent validities of the three questionnaires are 0.378 (SS), 0.708 (RP), and 0.463 (RDB), respectively, which demonstrates that the observed variables could be highly explained by the latent variables in the RPQ, while the interpretation proportion of observed variables in the SS is slightly worse.

Moreover, Table 6.4 reports the discriminant validity among potential variables in this study. A large correlation between potential variables means multiple potential variables jointly explaining one factor, which will lead to the problem of multi-collinearity. As shown in Table 6.4, it can be found that the square root of the Average Variance Extraction (AVE) of each potential variable is greater than the corresponding Pearson correlation, which means the correlation between potential variables is acceptable. Thus, each potential variable can represent different factors and the problem of multicollinearity would be not our concern.

6.1.4   METHODOLOGIES

After the measurement of drivers' sensation seeking, risk perception, and risky driving behaviors, specific statistical methods should be used to analyze the relationship between drivers' demographics, psychological factors, and risky driving behaviors. We have compared several commonly used statistical methods. After analyzing their advantages and disadvantages, we choose the method of SEM.

Most previous studies have used linear regression or multi-group correlation analysis in the relationships among demographics, traffic psychology, and driving safety. However, there are still some problems that cannot be solved through these methods. On the one hand, it is hard to measure the latent psychological and behavioral factors. On the other hand, it is also difficult to investigate complex relationships among driver's demographics, sensation seeking, risk perception, and risky driving behaviors in-depth. Previous studies have mainly focused on the effects of age or driving experience on personality traits associated with risky driving behaviors, or the effects of age or driving experience on risky driving behaviors solely, rather than incorporating them into a combined model.

Therefore, the method of the SEM is adopted in the current chapter. SEM is a tool that can handle many observed variables and specified latent variables by their linear combination and has been widely applied in various fields such as sociology, psychology, political science, as well as in traffic psychology and safety [156]. SEM consists of multiple measurement models and a structural model. The unobservable latent factors can be represented by multiple observable indicators in measurement models, while the structural model can explore the complex relationship among latent variables. Therefore, the SEM absorbs the advantages of many other methods (e.g., factor analysis and regression analysis) and has its unique advantages. Also, the SEM is generally used to investigate complex relationships among various endogenous variables and exogenous variables [157].

The SEM is composed of the structural model and measurement model. Measurement models use specific observed variables to characterize potential variables. The structural model explores the relationships between potential variables. Measurement models are normally specified in two sets of equations. The first set (the exogenous measurement model) is represented as follows:

$$X = \Lambda_X \xi + \delta, \tag{6.1}$$

where X denotes the vector of observed exogenous variables; Λ_X denotes the matrix of structural coefficients for latent exogenous variables to their observed indicator variables; ξ denotes the vector of latent exogenous constructs; and δ denotes the vector of measurement error terms for observed variables.

The second (endogenous measurement model) is a set of equations are summarized as follows:

$$Y = \Lambda_Y \eta + \varepsilon, \tag{6.2}$$

where Y denotes the vector of observed endogenous variables; and Λ_Y denotes the matrix of structural coefficients for latent endogenous variables to their observed indicator variables; and η denotes the vector of latent exogenous variables; and ε denotes the vector of measurement error terms for observed endogenous variables.

A structural model relating the exogenous latent variables and endogenous latent variables can be expressed as:

$$\eta = B\eta + \Gamma\xi + \zeta, \tag{6.3}$$

where η denotes the vector of the latent endogenous variable; B denotes the matrix of structural coefficients between endogenous latent variables; ξ denotes the vector of latent exogenous constructs; Γ denotes the matrix of structural coefficients for exogenous latent variables to endogenous latent variables; and ζ denotes the unexplainable part of latent variables contained in the model.

In this chapter, observed variables of the driver's age and driving years are denoted by X, observed variables of the questionnaires (SSS, RPQ, DBQ) after dimensionality reduction are denoted by Y. For latent variables, the driving experience is an exogenous latent variable represented by ξ. Psychological factors and driving behaviors are endogenous latent variables represented by η. A more detailed description of each variable in the SEM can be found in Table 6.5. The results of the SEM are obtained by AMOS® 26.0.

6.1.5   THE RELATIONSHIP AMONG DRIVERS' DRIVING EXPERIENCE, PSYCHOLOGICAL FACTORS, AND RISKY DRIVING BEHAVIORS

After the EFA and the CFA of those questionnaires, the SEM is then derived from the theoretical basis and proposed hypothesis described above. The structural model depicts the relationships among the driving experience, sensation seeking, risk perception, and risky driving behaviors. In this model, there are 13 observed variables and 4 latent variables in total. The standardized path coefficients of the SEM are shown in Figure 6.3.

To assess the goodness-of-fit of the hypothesized model, the χ^2 which measures the discrepancy between the hypothesized model and the data is used. An insignificant result indicates that the measurement model is consistent with the hypothesized model. However, the χ^2 value is very sensitive to the sample size, which is not a very practical fitness index. When the sample size is more than 200, the p-value of almost all studies is significant [158], which in our case are, $N = 3150, \chi^2(59) = 1162.364, \chi^2/df = 19.701, P < 0.001$. Instead, alternative global fitting indexes such as the Goodness of Fit Index (GFI), Adjusted Goodness of Fit Index ($AGFI$), Normative Fit Index (NFI), Comparative Fit Index (CFI), and the Tucker and Lewis Index

Table 6.5: The description of endogenous and exogenous variables in the structural model and measurement models

Latent Variables	Observed Variables	Description
Exogenous Variables		
ξ_1 (DE)		Driving experience.
	X_1 (Age)	Drivers' age.
	X_2 (Driving Years)	Drivers' driving years.
Endogenous Variables		
η_1 (SS)		Sensation seeking.
	Y_3 (Boredom)	Boredom susceptibility. A dislike of repetition of experience.
	Y_4 (Inhibition)	Disinhibition. The loss of social inhibitions.
	Y_5 (Experience)	Experience seeking. Its essence is "experience for its own sake."
	Y_6 (Thrill)	Thrill and adventure seeking. A desire to engage in outdoor sports or other activities.
η_2 (RP)		Risk perception.
	Y_7 (Violation)	The perception of taking deliberate illegal traffic rules of behavior.
	Y_8 (Aggressiveness)	The perception of engaging in aggressive driving behavior.
	Y_9 (Distraction)	The perception of taking the second driving task in the process of driving.
	Y_{10} (Careless)	The perception of careless driving habits.
η_3 (RDB)		Risky driving behaviors.
	Y_{11} (Error)	The departure of planned actions from some satisfactory path towards a desired goal.
	Y_{12} (Violation)	Deliberate deviations from those practices believed necessary to maintain the safe operation of a potentially hazardous system.
	Y_{13} (Aggressiveness)	Aggressive violation-closely related to the aggressive personality factors.
	Y_{14} (Lapse)	The unwitting deviation of action from the intention.

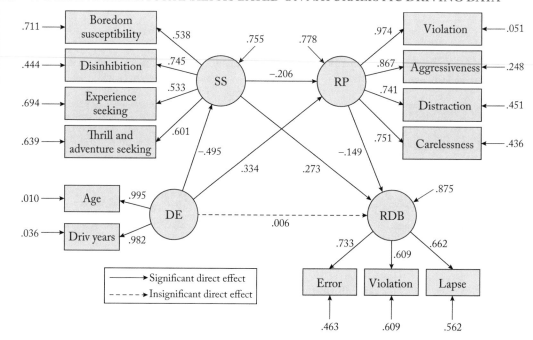

Figure 6.3: Standardized path coefficients of the SEM. *Note:* the dashed line indicates an insignificant direct effect, while the solid line indicates a significant effect; DE: driving experience; SS: sensation seeking; RP: risk perception; RDB: risky driving behaviors.

(*TLI*) are adopted. Contrary to the χ^2 goodness-of-fit, they are not easily affected by the sample size, which varies between 0 (no fit) to 1 (perfect fit). While values larger than 0.9 would generally indicate an adequate fit, which in our case are, $GFI = 0.943, AGFI = 0.913, NFI = 0.947, CFI = 0.950, TLI = 0.933$, respectively.

Also, the Root Mean Square Error of Approximation (*RMSEA*) and the Standardized Root Mean Square (*SRMR*) are included to assess the goodness-of-fit of the proposed model, which both vary between 0 (perfect fit) to 1 (no fit). While the RMSEA and SRMR values are less than 0.08 would generally indicate an adequate fit [159], which in our case are, $RMSEA = 0.076$ and $SRMR = 0.063$, respectively. Therefore, the goodness-of-fit of the hypothesized model is acceptable.

As shown in Figure 6.3 and Table 6.6, standardized regression weights, and the significance of the structural model can be acknowledged as follows:

1. The standardized factor loadings of driving experience on the SS and the RP are -0.495 ($p < 0.001$), 0.334 ($p < 0.001$), respectively, which indicates that driving experience has a significant positive effect on risk perception and a significant negative effect on sensation seeking.

Table 6.6: Regression coefficients of the structural model

Dependent	Independent	Estimate	S.E.	EST./S.E.	p-Value
SS	DE	−.495	.015	−32.923	***
RP	DE	.334	.018	18.250	***
	SS	−.206	.023	−8.876	***
RDB	DE	.006	.024	0.238	0.812
	SS	.273	.028	9.645	***
	RP	−.149	.026	−5.773	***

Note: ***: p-value < 0.001; S.E.: standard error, EST: estimate; DE: driving experience;
SS: sensation seeking; RP: risk perception; RDB: risky driving behaviors.

2. The standardized factor loading of the SS on the RDB is 0.273 ($p < 0.001$), and that of the RP on the RDB is −0.149 ($p < 0.001$), indicating that sensation seeking positively affects risky behaviors significantly, while risk perception negatively affects risky driving behaviors significantly.

3. Moreover, the standardized factor loading of the SS on the RP is −0.206 ($p < 0.001$), indicating that sensation seeking has a significant negative effect on risk perception.

4. Besides, the contribution of *age* and driving *years* can be obtained in the results of the SEM, which shows that both observed variables take up an almost equal proportion in the potential variable of driving experience (path regression estimations as 0.995 of driving experience on *age*, and 0.982 of driving experience on *driving years*).

In summary, according to the above analysis of the SEM, the direct effect of driving experience on risky driving behaviors is not significant ($p = 0.812$). However, the effect could be indirect, where driving experience affects risky driving behaviors through sensation seeking and risk perception. Table 6.7 shows the unstandardized result of the mediation model. The Bootstrap estimation method which resamples the data to obtain more accurate results is further adopted in the study, and estimations through Percentile and Bias Corrected are obtained.

As shown in Table 6.7, when sensation seeking and risk perception are included in the model as mediators, the total effect of driving experience on risky driving behaviors is significant ($p < 0.001$), where the regression weight decreased to −0.009. Moreover, all the indirect effects of driving experience on the RDB is significant while mediated by the SS (the regression weight as −0.006, $p < 0.001$), RP (the regression weight as −0.002, $p < 0.001$), respectively, or both of the SS and RP (the regression weight as −0.001, $p < 0.001$), which indicate the case of completely mediated effects. Therefore, driving experience affects risky driving behaviors completely and indirectly through sensation-seeking and risk perception.

Table 6.7: Mediating effects of driving experience on risky driving behaviors

Path	EST.	S.E.	EST./S.E.	p-Value	Lower 2.5%	Upper 2.5%	Lower 2.5%	Upper 2.5%
		Product of Coefficients			Percentile		Bias Corrected	
					Bootstrapping 1000 Times 95% CI			
Indirect Effects								
DE→SS→RDB	−.006	.001	−8.248	***	−.005	−.008	−.005	−.008
DE→RP→RDB	−.002	.000	−5.057	***	−.001	−.003	−.001	−.003
DE→SS→RP→RDB	−.001	.000	−4.235	***	.000	−.001	.000	−.001
Total indirect effects	−.009	.001	−10.624	***	−.008	−.011	−.008	−.011
Direct Effect								
DE→RDB	.000	.001	.203	0.839	−.002	.003	−.002	.003
Total Effect								
Total effects	−.009	.003	−7.824	***	−.007	−.011	−.007	−.011

Note: ***: p-value < 0.001; mediating factors: sensation seeking; risk perception; S.E.: standard error; EST: estimate; CI: confidence interval; DE: driving experience; SS: sensation seeking; RP: risk perception; RDB: risky driving behaviors.

Moreover, the mediating effect of sensation seeking on risky driving behaviors mediated by risk perception has also been found, which can be observed in Table 6.8. The direct effect of the SS on the RDB without mediation is significant (the regression weight as 0.059, $p < 0.001$). Moreover, when the RP is included as a mediator, the indirect effect of the SS on the RDB is also significant (the regression weight as 0.007, $p < 0.001$). And, the total effects of the SS on the RDB, including the direct and indirect effects, are significant (the regression weight as 0.065, $p < 0.001$). Therefore, the effect of sensation seeking on risky driving behaviors is a process of partial mediation, sensation seeking has a significant direct effect on risky driving behaviors, as well as an indirect effect through risk perception.

Results show that the driver's driving experience, which is represented by age and cumulative driving years, negatively affects the risky driving behavior engagements. The result indicates that driving experience negatively affects risky driving behavior engagements, which are well consistent with previous age-related analyses [117, 160] because age plays an important role in the study while regarding driving experience.

We further analyze the reasons that driving experience inhibits risky driving behaviors through the mediation model. The results of the SEM show that driving experience seems to be a distinct factor in risky driving behaviors. More specifically, there is no significant direct influence of driving experience on risky driving behavior engagements, while there exist significant indirect influences through sensation seeking and risk perception. Among them, driving experience indeed directly affects sensation-seeking and risk perception that are related to risky driving behaviors. The current study indicates that driving experience negatively affects sensation-seeking

Table 6.8: Mediating effects of sensation seeking on risky driving behaviors

Path	Product of Coefficients				Bootstrapping 1000 Times 95% CI			
					Percentile		Bias Corrected	
	EST.	S.E.	EST./S.E.	p-Value	Lower 2.5%	Upper 2.5%	Lower 2.5%	Upper 2.5%
Indirect Effect								
SS→RP→RDB	.007	.002	4.249	***	.004	.010	.004	.010
Direct Effect								
SS→RDB	.059	.007	8.589	***	.045	.072	.046	.072
Total Effect								
Total effects	.065	.007	9.651	***	.052	.078	.052	.079

Note: ***: p-value < 0.001; p-value < 0.001; mediating factors: risk perception; S.E.: standard error; EST: estimate; CI: confidence interval; SS: sensation seeking; RP: risk perception; RDB: risky driving behaviors.

while positively affect risk perception. With the increase of driving experience, the driver has a lower tendency to seek stimulation, which is consistent with the conclusion made by Nordfjærn et al. [161], in which lower sensation-seeking scores are associated with an incremental in age. On the contrary, an experienced driver has a higher ability to identify the severity of risky driving behaviors, which is also verified by Delhomme et al. [162].

Meanwhile, sensation seeking and risk perception are both highly correlated with risky driving behavior engagements. Specifically, the results in this study indicate that the driver's high sensation-seeking will promote to pursue risky driving behaviors, the stronger the driver's stimulation needs, the higher the frequency of engaging in risky driving behaviors. Conversely, with the improvement of risk perception, drivers have a lower probability to commit risky driving behaviors [163, 164]. In summary, certain psychological factors of experienced drivers tend to improve driving safety such as lower sensation seeking and higher risk perception, as a consequence, the likelihood of drivers' risky driving behavior engagements tend to be inhibited.

Furthermore, we have verified that sensation seeking and risk perception are potential mediators in the influence of driving experience on risky driving behaviors in the mediation model. Moreover, just the same as the results obtained from the SEM, driving experience seems not to have a significant direct effect on risky driving behaviors when ignoring the gender factor. The driving experience affects risky driving behaviors indirectly through sensation seeking and risk perception, which indicates a completely mediating effect. These strong indirect effects might explain the reason why the direct influence of driving experience on driving behaviors in the SEM is weak when adopting sensation seeking and risk perception as mediators.

Moreover, we also verify the relationship between sensation seeking and risk perception. The mediation model of sensation seeking on risky driving behaviors as shown in Table 6.8 indicates that sensation seeking negatively affects risk perception. Meanwhile, lower risk perception will further increase the likelihood of risky driving behavior engagements. Moreover, sensation seeking is a remote influence factor for risky driving behaviors, and risk perception plays a mediating role in the influence of sensation seeking on risky driving behaviors. Our results suggest that sensation seeking significantly affects risky driving behaviors both directly and indirectly, while risk perception mediates this kind of influence, which is also consistent with conclusions obtained by Ulleberg and Rundmo [127].

6.1.6   THE MODERATING RELATIONSHIP BETWEEN DRIVERS' CHARACTERISTICS AND RISKY DRIVING BEHAVIORS

The moderated mediation model of driving experience on risky driving behaviors regarding gender as an interaction factor is further analyzed. The different effects of driving experience on risky driving behaviors between male and female drivers are shown in Table 6.9 and Figure 6.4.

It can be observed that the total effect of the driving experience on the RDB (including direct and indirect effects) are significant both for the male drivers (the regression weight as -0.005, $p < 0.001$) and the female drivers (the regression weight as -0.012, $p < 0.001$), respectively, which suggests that the development trend of risky driving behavior engagements decrease with the increase of driving experience regardless of gender. And, the total effect for female drivers is significantly larger than male drivers ($p < 0.001$). However, the difference of total indirect effect is insignificant ($p = 0.153$), which indicates the indirect effects of driving experience on risky driving behaviors mediated by sensation seeking and risk perception are insignificantly distinct between different genders.

Moreover, the direct effect of the DE on the RDB is significant for male drivers (the regression weight is 0.005, $p < 0.05$) while insignificant for female drivers ($p = 0.970$), and their direct effect difference is marginally significant between different genders ($p = 0.066$). The result shows that, besides the negative indirect impact of driving experience on risky driving behaviors, the direct effect is significant and positive for male drivers. We may infer other undiscussed potential factors have positive effects on risky driving behaviors, which will be further addressed in later discussions. On the contrary, the insignificance of the direct effect for female drivers indicates that the negative impact of driving experience on risky behaviors could be completely mediated by sensation seeking and risk perception according to the current model. In brief, there exists a significant difference in the influence trends of driving experience on risky driving behaviors between different genders.

When analyzing the moderating role of gender, the total effects of driving experience on risky driving behavior engagements are negative and significant both in male and female drivers. However, the influence trends of driving experience on risky driving behavior engagements are distinct between genders, in which the development trend of risky driving behavior engagements for female drivers declines greater with the growth of driving experience. The main reason for the difference between genders in the current study is that, due to the mediation of sensation seek-

Table 6.9: Moderated mediation model of driving experience on risky driving behavior

| | Product of Coefficients | | | | Bootstrapping 1000 Times 95% CI | | | |
| | | | | | Percentile | | Bias Corrected | |
Path	EST.	S.E.	EST./S.E.	p-Value	Lower 2.5%	Upper 2.5%	Lower 2.5%	Upper 2.5%
Male Group								
Indirect Effects								
DE→SS→RDB	−.006	.001	−5.082	***	−.008	−.004	−.008	−.004
DE→RP→RDB	−.003	.001	−3.713	***	−.005	−.001	−.005	−.001
DE→SS→RP→RDB	−.001	.001	−2.639	.013	−.001	.000	−.001	.000
Total indirect effect	−.010	.001	−6.985	***	−.012	−.007	−.012	−.007
Direct Effect								
DE→RDB	.005	.002	2.438	.015	.001	.008	.001	.009
Total Effect								
Total effect	−.005	.001	−3.361	.001	−.008	−.002	−.008	−.002
Female Group								
Indirect Effects								
DE→→SS→RDB	−.009	.001	−7.695	***	−.011	−.007	−.011	−.007
DE→RP→RDB	−.003	.001	−3.758	***	−.004	−.001	−.004	−.001
DE→SS→RP→RDB	−.001	.000	−3.406	.001	−.001	.000	−.001	.000
Total indirect effect	−.012	.002	−8.020	***	−.015	−.010	−.015	−.010
Direct Effect								
DE→RDB	.000	.002	.038	.970	−.003	.004	−.003	.004
Total Effect								
Total effect	−.012	.003	−7.727	***	−.015	−.009	−.015	−.009
Difference Between Genders								
Total effect difference	.007	.002	3.562	***	.003	.011	.003	.011
Direct effect difference	.005	.003	1.840	.066	−.010	.001	−.010	.001
Indirect effect difference	.003	.002	1.429	.153	−.001	.006	−.001	.006

Note: ***: p-value < 0.001; moderating factor: gender; mediating factors: sensation seeking and risk perception; S.E.: standard error; EST: estimate; CI: confidence interval; DE: driving experience; SS: sensation seeking; RP: risk perception; RDB: risky driving behaviors.

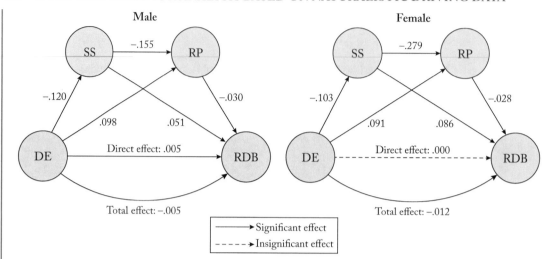

Figure 6.4: The direct, indirect, and total effects of driving experience on risky driving behaviors in male and female drivers. *Notes:* DE: driving experience; SS: sensation seeking; RP: risk perception; RDB: risky driving behaviors.

ing and risk perception, we found that the driving experience of the male drivers has a positive direct influence on risky driving behaviors, while there is no significant direct effect for female drivers. And there exists a marginally significant difference regarding the direct effects of driving experience on risky driving behaviors between genders. The positive direct effect for males partially offsets but does not completely conceal the negative indirect effects, thus, resulting in a weaker negative total effect than female drivers. In other words, the effects of driving experience on risky driving behaviors are partially mediated by sensation seeking and risk perception for male drivers, while completely mediated for female drivers in the current model.

It is still uncertain what factors account for the difference between genders, however, the reason might be related to the functional health of senior drivers which encompasses physical, cognitive, and psychomotor ability. Aging is accompanied by declines in cognitive, sensory, and motor abilities that can make the task of driving more difficult [165]. And according to the previous research based on the SHRP 2 database, senior drivers' poor functional health will promote the engagement of crash risk [166]. Generally, aged drivers are aware of functional decline and then avoid travel under threatening conditions [167], however, Bauer et al. have found that male drivers are less likely to reduce or stop driving under some adverse conditions than female drivers [168]. Moreover, self-regulation in driving is generally described as the process of modifying or adjusting one's driving patterns by driving less or intentionally avoiding challenging driving [169, 170] suggest that the degree of self-regulation both for the male and female drivers are increasing with the age, however, male drivers seem to have more confidence in driving, even in the condition of poor functional health, which may cause that the degree of self-regulation is less increased with age when compared with female drivers. Therefore, poor functional health, less self-regulation, or the lower tendency to reduce driving might contribute to more engagements of risky driving behaviors for senior male drivers other than senior female drivers. Considering that the age factor is also an important contributor to the driving experience in our study, thus, it could be reflected in the male drivers' positive direct effect of driving experience on risky driving behaviors as indicated in this study.

6.2   CLASSIFICATION OF DRIVER'S DRIVING RISK BY RANDOM FORREST ALGORITHM

Previous studies have used a logistic regression method to identify the risk level of drivers based on driver's personality traits, age, and critical-incident events (CIE) rate of the 100-Car Naturalistic Driving Study [171]. However, we have more accurate methods to identify the level of risk for drivers at present. In this chapter, we constructed a classification model of driver risk to explore the classification performance of the self-reported driver characteristics and the driving behaviors by using the RF method. The CNC rate is clustered into three risk levels by the K-means clustering method, and then the clustering result is used as the label of the driver risk level in this classification model. Then, drivers' demographics (driving experience, gender), psychological factors (sensation seeking, risk perception), and the DBQ are included as predictive factors in the RF classifier. Thus, the driving risk level of drivers is expected to be identified in the RF classifier.

6.2.1   METHODOLOGIES

The K-means clustering and the RF classifier are, respectively, popular unsupervised and supervised machine learning methods. The cluster analysis provides an objective approach to classifying drivers into different risk levels and has been used in traffic safety research. In this paper, the K-means clustering method is adopted to obtain the probability distribution of the CNC rate data. The advantage is that the K-means clustering is an unsupervised learning method that does not require training labels and it is very flexible to select the different numbers of clusters.

First, K types of samples predefined before clustering are randomly selected as the clustering center. Then, each data is classified into the cluster whose mean is closest to the data value. To ensure the optimal allocation of data, the optimization index of the K-means method, called within-cluster sum of squares is widely used

$$J = \sum_{k=1}^{k} \sum_{i=1}^{n} ||x_i - u_k||^2, \tag{6.4}$$

where x_i denotes the data which is represented by CNC in the current study and u_k denotes the preset "center" of each cluster. The allocation process is to minimize J.

The sample mean of each cluster will be calculated as the new clustering center. The algorithm will end if the cluster center no longer changes or the number of iterations reached. Finally, three clusters in total are generated by the K-means clustering, representing the low-, middle-, and high-risk levels of drivers, respectively.

To identify the driver risk, a supervised machine learning approach is adopted. The RF algorithm achieved the best results in a large number of classification algorithms. Thus, in the current chapter, the RF classifier is used to identify the driver risk through drivers' demographics, psychological factors, and risky driving behaviors.

The RF is an ensemble learning method combined with multiple weighted decision trees which is a weak learner and has limited ability to handle large samples and multiple dimensions. The decision tree is a popular machine learning method that has been used in many pattern recognition and classification tasks [172]. A decision tree is constructed with one root node, multiple middle nodes, and leaf nodes whose results are the basis for classification. Specifically, the algorithm of Classification and Regression Tree (CART) involved in this study uses the Gini coefficient instead of the information gain ratio. Gini coefficient represents the impurity of the model, and the smaller Gini coefficient, the lower impurity, and the better feature. Gini coefficient expression of the probability distribution is:

$$Gini(p) = \sum_{k=1}^{n} p_k(1 - p_k) = 1 - \sum_{k=1}^{n} p_k^2, \qquad (6.5)$$

where k is the categories of the model (three in the study) and p_k is the probability of the k_{th} category in the sample.

Suppose that the number of sample D is $|D|$. The Gini coefficient expression of sample D can be obtained:

$$Gini(D) = 1 - \sum_{k=1}^{n} \left(\frac{|C_k|}{|D|} \right)^2, \qquad (6.6)$$

where C_k denotes the number of kth category in the sample.

The classification will start at the root node which represents the entire sample. At each middle node, a feature with the minimum Gini index is preferentially selected to partition the sample. By traversing all values of feature A as the partition node, sample D is divided into $|D_1|$ and $|D_2|$ each time. Then, under the condition of feature A, the Gini coefficient expression of sample D is as follows:

$$Gini(D, A) = \frac{|D_1|}{|D|} Gini(D_1) + \frac{|D_2|}{|D|} Gini(D_2). \qquad (6.7)$$

The value corresponding to the minimum $Gini$ coefficient of feature A will be used to partition samples at the current middle node. Subsequent nodes will remove this feature and repeat to select another feature with minimum $Gini$ index until the categories in the dataset are the same, or the set threshold is reached. Finally, the RF combines multiple random sampled decision trees to reduce the risk of overfitting and enhance generalization ability [173].

In the current chapter, the RF classification is used to identify driver risk, and predictive factors include drivers' self-reported demographics, psychological factors, and risk driving behaviors. The K-means clustering and the RF algorithm are analyzed with the help of Python 3.

6.2.2   CLUSTERING OF DRIVER'S RISK DEGREE

Since the RF is a supervised learning method, we need to label the risk level of the driver before training the model. Therefore, we need an algorithm to identify the risk level of each driver in advance. One possible way to identify high-risk drivers is to detect a crash involving kinetic energy transfer or dissipation. However, the crash is a small probability event in daily driving, so it is difficult to distinguish the risk level of drivers through the crash and find its correlation with driver's characteristics and behaviors. Researchers need alternative indicators of driver risk with a higher frequency of occurrence. As described by Bagdadi, there is a correlation between the frequency of critical braking events and crashes [174]. Similarly, the speed at the beginning of braking had a strong relationship with near-crash events [175]. Therefore, the near-crash evaluated by high breaking acceleration and low TTC is developed to solve the problem of low collision occurrences. Near-crash is defined as "a conflict situation that requires a rapid, severe evasive maneuver to avoid a crash" and has been proven to have a correlation and similar kinematic characteristics with the crash [176].

Compared with self-reported collision events, the usage of the CNC of naturalistic driving data has two advantages. First, the CNC of naturalistic driving data is more accurate than self-reported collision records because it is not generally affected by the driver's subjective evaluation criteria. Second, video records of naturalistic driving data can be repeatedly verified and analyzed. Such that, the CNC is widely used in the research of driving safety [177]. Therefore, the CNC rate as an indicator to evaluate the driver's driving style is adopted in the chapter.

The CNC rates obtained from the SHRP 2 naturalistic driving data are clustered into three risk levels by the K-mean cluster method. As shown in Figure 6.5 of the clustering results, the "low-risk" cluster is of the rate below 0.572, the "middle-risk" cluster with the rate ranging from 0.583–1.813, and the "high-risk" cluster ranges from 1.834–10.333.

The driving experience, gender, sensation seeking, risk perception, and risky driving behaviors are selected as predictive factors in this classification model, which are obtained from the results of the SEM to offer adequate explanatory abilities for the classification model. On the other way around, the result of the classification model can also verify the validity of the SEM. Eventually, a total of 14 predictive variables (*Age*, *Driving years* of driving experience; *Boredom*, *Inhibition*, *Experience*, *Thrill* of sensation seeking; *Aggressiveness*, *Violation*, *Distraction*, *Careless* of risk perception; *Violation*, *Error*, *Lapse* of risky driving behaviors; gender) are utilized in this classification model, which is all normalized to ensure the same dimension before the training.

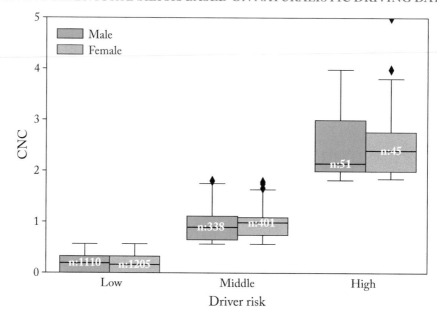

Figure 6.5: The boxplot of clustering distribution of the driver risk by using *K*-mean cluster. *Note:* clustering variable: the CNC rate.

Moreover, according to the clustering result shown in Table 6.10 and Figure 6.5, the quantity of low-risk clusters is much larger than other groups, taking up 73.5% of the overall sample, while middle-risk and high-risk clusters share the rest data with a proportion of 23.5% and 3.0%, respectively. It can be noted that there is indeed a difference in the average age and driving age of different risk groups, and the age and driving years of high-risk drivers are lower than that of low-risk drivers, while there is no significant gender difference. To maintain a reasonable balance between these proportions, as well as the diversity of clusters, the method of the Synthetic Minority Oversampling Technique (SMOTE) is adopted to middle and high clusters. Thus, the final number of the sample included in the RF classifier is 6945 (2315 samples of low-risk, medium-risk, and high-risk levels, respectively).

6.2.3   CLASSIFICATION OF DRIVER'S RISK DEGREE

The risk levels of drivers are further utilized as the label in the RF classifier. After training the model, the cross-validation method is adopted in the RF classifier, and the sample is cross-validated with 10 folds to achieve more than 90.8% average accuracy. The detailed classification performance in the RF classifier is illustrated by a confusion matrix as shown in Figure 6.6. According to the result, the low-risk level achieves up to 90.8% accuracy, while the remaining 9% of low-risk drivers were classified as the middle-risky drivers incorrectly. The classification

Table 6.10: Characteristics of driver risk groups

Risk of Groups	Number of Drivers	Mean CNC Rate	% of Males in Each Group	Mean Driving Years	Mean Age
Low	2315	0.191	47.9	26.972	40–44
Middle	739	0.965	45.7	18.8	35–39
High	96	2.688	53.1	18.1	30–34

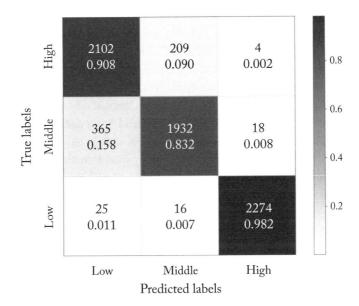

Figure 6.6: The confusion matrix of classification result using the Random Forest. *Note: N =* 6945; *K*-fold = 10; predictive variables: driving experience, gender, sensation seeking, risk perception, and dangerous driving behaviors.

result of the middle-risk level is of the least satisfactory, with 83.5% of samples correctly classified and 15.8% of remaining samples incorrectly classified to the low-risk level. The classification of high-risk drivers is the best, which achieves an accuracy of up to 98.2%. It can be seen that the classification ability of high-risk drivers is very strong, which is hard to be confused with that of low and middle-risk drivers, while a small part of low and middle-risk drivers are confused with each other. However, the overall classification effect is satisfactory.

6.2.4   ANALYSIS OF CLASSIFICATION MODEL

The result of the clustering model shows that individual driver risk varies substantially with three distinct risk-level groups. The cluster analysis indicates that the CNC rate of the higher risk drivers is over 10 times greater than that of the low-risk groups, and the CNC rate of the moderate risk drivers is over 4 times greater than that of the low-risk group, which is consistent with the study of Guo and Fang et al. [171]. The accuracy of the classification of high-risk drivers achieves up to 98.2%, which is hard to be confused with low- and middle-risk drivers.

The results of the classification model suggest that the driver risk can be assessed by self-reported driver's demographics, sensation seeking, risk perception, and risky driving behaviors, instead of the analysis of the driver's daily driving data. Although high-risk drivers only account for a small proportion of the driver population, they have a substantial impact on overall traffic safety. Identifying factors associated with individual driving risk, detecting unsafe driving behaviors, and identifying high-risk drivers will enable proper driver-behavior intervention, as well as safety countermeasures to reduce the crash likelihood of high-risk groups and improve overall driving safety. After all, it is unrealistic to monitor all collision accidents in daily driving, screen out high-risk drivers to have them guided and supervised. On the one hand, the collision is generally considered as a small probability event that is hard to be supervised. On the other hand, measures-taking after an accident is usually based on the loss of property and damage to health. If specific characteristics of drivers can be used to identify the high-risk drivers and have them supervised and guided before the occurrence of accidents, it will be beneficial to public traffic safety.

Identification of the driver's driving risk level might be also helpful for the strategy development of collaborative driving or shared driving. So far, few pieces of existing research on collaborative driving strategies take drivers at different risk levels into account. Nevertheless, different characteristics of drivers might lead to various driving styles, such that personal cooperative driving strategies are required to be particularly designed. To avoid serious potential conflicts between the decision-making system of the intelligent vehicle and the driver himself/herself, it is promising to deepen the research of incorporating driving styles into the shared driving strategies to improve driving safety and driving comfort in the future.

6.3   SUMMARY

This chapter verified the impact of driving experience on sensation seeking, risk perception, and risky driving behaviors, while also analyzing the mediating role of sensation seeking and risk perception, as well as the moderating role of gender by using the SEM, which is based on a large-scale questionnaire survey from the SHRP 2. In summary, the main conclusions of this chapter are summarized as follows.

1. Drivers' driving experience negatively affects sensation seeking, while positively affect risk perception.

2. Sensation seeking and risk perception affect risky driving behaviors of drivers, lower sensation-seeking, and higher risk perception inhibit the likelihood of risky driving behavior engagements.

3. Drivers' driving experience indirectly affects the likelihood of risky driving behavior engagements, regardless of gender, in which sensation seeking and risk perception mediate the negative effects of driving experience on risky driving behaviors.

4. There exists a significant distinction in the influence trends of driving experience on risky driving behaviors between different genders, in which the development trend of risky driving behavior engagements for female drivers declines greater with the growth of driving experience.

In summary, these findings suggest the necessity to embed issues concerning specific aspects of sensation seeking and risk perception into the planning of road safety campaigns and policies, and safety policies should pay more attention to young drivers with less driving experience, higher sensation seeking, and lower risk perception. The traditional strategy of traffic safety campaigns was using authority to tell young drivers to drive safely. But, as an alternative measure to law and authority, it let drivers be aware of the need for behavioral change on their initiative could be more effective [127]. Also, video and simulator training can significantly improve driving safety [7], such that developing safety education and publicity through new media to reduce the drivers' pursuit of driving stimulation, and enhance drivers' cognitive ability for driving risks might be more effective. Further, drivers who scored higher in sensation seeking and lower in risk perception should be guided to participate in a simulation training program to improve their driving experience and awareness of driving risk, which might include, but is not limited to, evaluating driving ability, handling high-risk driving scenarios, and training driving skill.

Besides, the driving risk level of drivers evaluated by the CNC rate could be classified through drivers' self-reported questionnaires of demographics, sensation seeking, risk perception, and risky driving behaviors. Identifying potential high-risk drivers will be beneficial to provide more targeted safety education and driving guidance for these drivers, and ultimately, reduce the crash likelihood of high-risk groups and improve overall driving safety.

Bibliography

[1] World Health Organization. (2018). Global status report on road safety 2018. World Health Organization. Geneva, WHO. https://www.who.int/violence_injury_prevention/road_safety_status/2018/en/ 1

[2] Lawton, R., Parker, D., Manstead, A. S., and Stradling, S. G. (1997). The role of affect in predicting social behaviors: The case of road traffic violations. *Journal of Applied Social Psychology*, 27(14):1258–1276. DOI: 10.1111/j.1559-1816.1997.tb01805.x. 1

[3] Singh, S. (2015). Critical reasons for crashes investigated in the national motor vehicle crash causation survey (no. DOT HS 812 115). 1

[4] SAE International (2014). Taxonomy and definitions for terms related to on-road motor vehicle automated driving systems. *SAE Journal*, pages 3016–2014. DOI: j3016_201401. 1

[5] Ma, M., Yan, X., Huang, H., and Abdel-Aty, M. (2010). Safety of public transportation occupational drivers: Risk perception, attitudes, and driving behavior. *Transportation Research Record*, 2145(1):72–79. DOI: 10.3141/2145-09. 1

[6] Ram, T. and Chand, K. (2016). Effect of drivers' risk perception and perception of driving tasks on road safety attitude. *Transportation Research Part F: Traffic Psychology and Behaviour*, 42:162–176. DOI: 10.1016/j.trf.2016.07.012. 1

[7] Zhao, X., Xu, W., Ma, J., Li, H., and Chen, Y. (2019). An analysis of the relationship between driver characteristics and driving safety using structural equation models. *Transportation Research Part F: Traffic Psychology and Behaviour*, 62:529–545. DOI: 10.1016/j.trf.2019.02.004. 1, 139

[8] Panayiotou, G. (2015). The bold and the fearless among us: Elevated psychopathic traits and levels of anxiety and fear are associated with specific aberrant driving behaviors. *Accident Analysis and Prevention*, 79:117–125. DOI: 10.1016/j.aap.2015.03.007. 2

[9] Bogdan, S. R., Măirean, C., and Havârneanu, C. E. (2016). A meta-analysis of the association between anger and aggressive driving. *Transportation Research Part F: Traffic Psychology and Behaviour*, 42:350–364. DOI: 10.1016/j.trf.2016.05.009. 2

[10] Rajamani, R. (2011). *Vehicle Dynamics and Control*. Springer Science and Business Media. DOI: 10.1007/978-1-4614-1433-9. 2

[11] Lefèvre, S., Vasquez, D., and Laugier, C. (2014). A survey on motion prediction and risk assessment for intelligent vehicles. *ROBOMECH Journal*, 1(1):1–14. DOI: 10.1186/s40648-014-0001-z. 3

[12] Brännström, M., Coelingh, E., and Sjöberg, J. (2010). Model-based threat assessment for avoiding arbitrary vehicle collisions. *IEEE Transactions on Intelligent Transportation Systems*, 11(3):658–669. DOI: 10.1109/tits.2010.2048314. 3, 4

[13] Lin, C. F., Ulsoy, A. G., and LeBlanc, D. J. (2000). Vehicle dynamics and external disturbance estimation for vehicle path prediction. *IEEE Transactions on Control Systems Technology*, 8(3):508–518. DOI: 10.1109/87.845881. 3

[14] Huang, J. and Tan, H. S. (2006, June). Vehicle future trajectory prediction with a DGPS/INS-based positioning system. *American Control Conference*, page 6, IEEE. DOI: 10.1109/acc.2006.1657655. 3

[15] Pepy, R., Lambert, A., and Mounier, H. (2006, June). Reducing navigation errors by planning with realistic vehicle model. *IEEE Intelligent Vehicles Symposium*, pages 300–307, IEEE. DOI: 10.1109/ivs.2006.1689645. 3

[16] Eidehall, A. and Petersson, L. (2008). Statistical threat assessment for general road scenes using Monte Carlo sampling. *IEEE Transactions on Intelligent Transportation Systems*, 9(1):137–147. DOI: 10.1109/tits.2007.909241. 3

[17] Kaempchen, N., Schiele, B., and Dietmayer, K. (2009). Situation assessment of an autonomous emergency brake for arbitrary vehicle-to-vehicle collision scenarios. *IEEE Transactions on Intelligent Transportation Systems*, 10(4):678–687. DOI: 10.1109/tits.2009.2026452. 3

[18] Schubert, R., Richter, E., and Wanielik, G. (2008). Comparison and evaluation of advanced motion models for vehicle tracking. *11th International Conference on Information Fusion*, pages 1–6, IEEE. 3

[19] Ammoun, S. and Nashashibi, F. (2009). Real time trajectory prediction for collision risk estimation between vehicles. *5th International Conference on Intelligent Computer Communication and Processing*, pages 417–422, IEEE. DOI: 10.1109/iccp.2009.5284727. 3

[20] Kaempchen, N., Weiss, K., Schaefer, M., et al. (2004). IMM object tracking for high dynamic driving maneuvers. *IEEE Intelligent Vehicles Symposium*, pages 825–830, IEEE. DOI: 10.1109/ivs.2004.1336491. 3

[21] Hillenbrand, J., Spieker, A. M., and Kroschel, K. (2006). A multilevel collision mitigation approach—Its situation assessment, decision making, and performance tradeoffs. *IEEE Transactions on Intelligent Transportation Systems*, 7(4):528–540. DOI: 10.1109/TITS.2006.883115. 3, 4

[22] Polychronopoulos, A., Tsogas, M., Amditis, A. J., et al. (2007). Sensor fusion for predicting vehicles' path for collision avoidance systems. *IEEE Transactions on Intelligent Transportation Systems*, 8(3):549–562. DOI: 10.1109/TITS.2007.903439. 3

[23] Miller, R. and Huang, Q. (2002). An adaptive peer-to-peer collision warning system. *Vehicular Technology Conference. IEEE 55th Vehicular Technology Conference*. 1:317–321, VTC Spring, IEEE. DOI: 10.1109/vtc.2002.1002718. 3, 4

[24] Barth, A. and Franke, U. (2008). Where will the oncoming vehicle be the next second? *IEEE Intelligent Vehicles Symposium*, pages 1068–1073. DOI: 10.1109/ivs.2008.4621210. 3

[25] Tan, H. S. and Huang, J. (2006). DGPS-based vehicle-to-vehicle cooperative collision warning: Engineering feasibility viewpoints. *IEEE Transactions on Intelligent Transportation Systems*, 7(4):415–428. DOI: 10.1109/tits.2006.883938. 3

[26] Batz, T., Watson, K., and Beyerer, J. (2009). Recognition of dangerous situations within a cooperative group of vehicles. *IEEE Intelligent Vehicles Symposium*, pages 907–912. DOI: 10.1109/ivs.2009.5164400. 3

[27] Lytrivis, P., Thomaidis, G., and Amditis, A. (2008). Cooperative path prediction in vehicular environments. *11th International IEEE Conference on Intelligent Transportation Systems*, pages 803–808. DOI: 10.1109/itsc.2008.4732629. 3, 4

[28] Bishop, G. and Welch, G. (2001). An introduction to the Kalman filter. *Proc of SIGGRAPH*, Course 8, 41:27599–23175. 4

[29] Veeraraghavan, H., Papanikolopoulos, N., and Schrater, P. (2006). Deterministic sampling-based switching Kalman filtering for vehicle tracking. *IEEE Intelligent Transportation Systems Conference*, pages 1340–1345. DOI: 10.1109/itsc.2006.1707409.

[30] Dyckmanns, H., Matthaei, R., Maurer, M., et al. (2011). Object tracking in urban intersections based on active use of a priori knowledge: Active interacting multi model filter. *IEEE Intelligent Vehicles Symposium (IV)*, pages 625–630. DOI: 10.1109/ivs.2011.5940443.

[31] Christopher, T. (2009). Analysis of dynamic scenes: Application to driving assistance. *Institut National Polytechnique de Grenoble-INPG*. 5

[32] Joseph, J., Doshi-Velez, F., Huang, A. S., et al. (2011). A Bayesian nonparametric approach to modeling motion patterns. *Autonomous Robots*, 31(4):383–400. DOI: 10.1007/s10514-011-9248-x. 5

[33] Aoude, G., Joseph, J., Roy, N., et al. (2011). Mobile agent trajectory prediction using Bayesian nonparametric reachability trees. *Infotech@ Aerospace*, page 1512. DOI: 10.2514/6.2011-1512. 4, 5

[34] Tran, Q. and Firl, J. (2014). Online maneuver recognition and multimodal trajectory prediction for intersection assistance using non-parametric regression. *IEEE Intelligent Vehicles Symposium Proceedings*, pages 918–923. DOI: 10.1109/ivs.2014.6856480. 4, 5

[35] Hu, W., Xiao, X., Fu, Z., et al. (2006). A system for learning statistical motion patterns. *IEEE Transactions on Pattern Analysis and Machine Intelligence*, 28(9):1450–1464. DOI: 10.1109/tpami.2006.176. 5

[36] Atev, S., Miller, G., and Papanikolopoulos, N. P. (2010). Clustering of vehicle trajectories. *IEEE Transactions on Intelligent Transportation Systems*, 11(3):647–657. DOI: 10.1109/TITS.2010.2048101. 5

[37] Buzan, D., Sclaroff, S., and Kollios, G. (2004). Extraction and clustering of motion trajectories in video. *Proc. of the 17th International Conference on Pattern Recognition, ICPR*, 2:521–524, IEEE. DOI: 10.1109/icpr.2004.1334287. 4, 5

[38] Hermes, C., Wohler, C., Schenk, K., et al. (2009). Long-term vehicle motion prediction. *IEEE Intelligent Vehicles Symposium*, pages 652–657. DOI: 10.1109/ivs.2009.5164354. 4, 5

[39] Schlechtriemen, J., Wedel, A., Hillenbrand, J., et al. (2014). A lane change detection approach using feature ranking with maximized predictive power. *IEEE Intelligent Vehicles Symposium Proceedings*, pages 108–114. DOI: 10.1109/ivs.2014.6856491. 5

[40] Geng, X., Liang, H., Yu, B., et al. (2017). A scenario-adaptive driving behavior prediction approach to urban autonomous driving. *Applied Sciences*, 7(4):426. DOI: 10.3390/app7040426. 5

[41] Hu, M., Liao, Y., Wang, W., et al. (2017). Decision tree-based maneuver prediction for driver rear-end risk-avoidance behaviors in cut-in scenarios. *Journal of Advanced Transportation*, 7170358:1–12. DOI: 10.1155/2017/7170358. 5

[42] Huang, R., Liang, H., Zhao, P., et al. (2017). Intent-estimation-and motion-model-based collision avoidance method for autonomous vehicles in urban environments. *Applied Sciences*, 7(5):457. DOI: 10.3390/app7050457. 5

[43] Tamke, A., Dang, T., and Breuel, G. (2011). A flexible method for criticality assessment in driver assistance systems. *IEEE Intelligent Vehicles Symposium (IV)*, pages 697–702. DOI: 10.1109/ivs.2011.5940482. 6

[44] Houenou, A., Bonnifait, P., Cherfaoui, V., et al. (2013). Vehicle trajectory prediction based on motion model and maneuver recognition. *IEEE/RSJ International Conference on Intelligent Robots and Systems*, pages 4363–4369. DOI: 10.1109/iros.2013.6696982. 6

[45] Aoude, G., Luders, B., Lee, K., et al. (2010). Threat assessment design for driver assistance system at intersections. *13th International IEEE Conference on Intelligent Transportation Systems*, pages 1855–1862. DOI: 10.1109/itsc.2010.5625287. 6

[46] Käfer, E., Hermes, C., Wöhler, C., et al. (2010). Recognition of situation classes at road intersections. *IEEE International Conference on Robotics and Automation*, pages 3960–3965. DOI: 10.1109/robot.2010.5509919. 6

[47] Lawitzky, A., Althoff, D., Passenberg, C. F., et al. (2013). Interactive scene prediction for automotive applications. *IEEE Intelligent Vehicles Symposium (IV)*, pages 1028–1033. DOI: 10.1109/ivs.2013.6629601. 6

[48] Brand, M., Oliver, N., and Pentland, A. (1997). Coupled hidden Markov models for complex action recognition. *Proc. of IEEE Computer Society Conference on Computer Vision and Pattern Recognition*, pages 994–999. DOI: 10.1109/cvpr.1997.609450. 6

[49] Agamennoni, G., Nieto, J. I., and Nebot, E. M. (2012). Estimation of multivehicle dynamics by considering contextual information. *IEEE Transactions on Robotics*, 28(4):855–870. DOI: 10.1109/tro.2012.2195829. 6

[50] Agamennoni, G., Nieto, J. I., and Nebot, E. M. (2011). A Bayesian approach for driving behavior inference. *IEEE Intelligent Vehicles Symposium (IV)*, pages 595–600. DOI: 10.1109/ivs.2011.5940407. 6

[51] Jianping, L. (2018). Research on prediction method of traffic vehicle motion for intelligent driving. Jilin University, Changchun, PRC. 6

[52] Talebpour, A., Mahmassani, H. S., and Hamdar, S. H. (2015). Modeling lane-changing behavior in a connected environment: A game theory approach. *Transportation Research Procedia*, 7:420–440. DOI: 10.1016/j.trpro.2015.06.022. 6

[53] Oyler, D. W., Yildiz, Y., Girard, A. R., et al. (2016). A game theoretical model of traffic with multiple interacting drivers for use in autonomous vehicle development. *American Control Conference (ACC)*, pages 1705–1710, IEEE. DOI: 10.1109/acc.2016.7525162. 6

[54] Alahi, A., Goel, K., Ramanathan, V., et al. (2016). Social LSTM: Human trajectory prediction in crowded spaces. *Proc. of the IEEE Conference on Computer Vision and Pattern Recognition*, pages 961–971. DOI: 10.1109/cvpr.2016.110. 6

[55] Deo, N. and Trivedi, M. M. (2018). Convolutional social pooling for vehicle trajectory prediction. *Proc. of the IEEE Conference on Computer Vision and Pattern Recognition Workshops*, pages 1468–1476. DOI: 10.1109/cvprw.2018.00196. 6

[56] Xuewu, J., Cong, F., Xiangkun, H., et al. (2019). Driving intention recognition and vehicle trajectory prediction based on LSTM network. *Journal of China Highway*, 32(6):34–42. 6

[57] Quintero, R., Parra, I., Llorca, D. F., et al. (2015). Pedestrian intention and pose prediction through dynamical models and behaviour classification. *IEEE International Conference on Intelligent Transportation Systems*. DOI: 10.1109/itsc.2015.22. 7, 8

[58] Volz, B., Behrendt, K., Mielenz, H., et al. (2016). A data-driven approach for pedestrian intention estimation. *IEEE 19th International Conference on Intelligent Transportation Systems (ITSC)*. DOI: 10.1109/itsc.2016.7795975. 7

[59] Koehler, S. (2015). Stereo-vision-based pedestrian's intention detection in a moving vehicle. *IEEE 18th International Conference on Intelligent Transportation Systems*. DOI: 10.1109/itsc.2015.374. 7

[60] Fang, Z., David, V., and Antonio, L. (2017). On-board detection of pedestrian intentions. *Sensors*, 17(10):2193. DOI: 10.3390/s17102193. 7

[61] Ghori, O., Mackowiak, R., Bautista, M., et al. (2018). Learning to forecast pedestrian intention from pose dynamics. *IEEE Intelligent Vehicles Symposium (IV)*. DOI: 10.1109/ivs.2018.8500657. 8

[62] Schulz, A. T. and Stiefelhagen, R. (2015). Pedestrian intention recognition using latent-dynamic conditional random fields. *Intelligent Vehicles Symposium (IV)*, IEEE. DOI: 10.1109/ivs.2015.7225754. 8

[63] Schneider, N. and Gavrila, D. M. (2013). Pedestrian path prediction with recursive Bayesian filters: A comparative study. *German Conference on Pattern Recognition*, Springer, Berlin, Heidelberg. DOI: 10.1007/978-3-642-40602-7_18. 8, 61

[64] Keller, C. G. and Gavrila, D. M. (2014). Will the pedestrian cross? A study on pedestrian path prediction. *IEEE Transactions on Intelligent Transportation Systems*, 15(2):494–506. DOI: 10.1109/tits.2013.2280766. 8

[65] Lee, N., Choi, W., Vernaza, P., Choy, C. B., Torr, P. H., and Chandraker, M. (2017). Desire: Distant future prediction in dynamic scenes with interacting agents. *Proc. of the IEEE Conference on Computer Vision and Pattern Recognition*, pages 336–345. DOI: 10.1109/cvpr.2017.233. 8

[66] Hug, R., Becker, S., Hübner, W., et al. (2018). Particle-based pedestrian path prediction using LSTM-MDL models. *21st International Conference on Intelligent Transportation Systems (ITSC)*, IEEE. DOI: 10.1109/itsc.2018.8569478. 8

[67] Saleh, K., Hossny, M., and Nahavandi, S. (2017). Intent prediction of vulnerable road users from motion trajectories using stacked LSTM network. *IEEE 20th International Conference on Intelligent Transportation Systems (ITSC)*. DOI: 10.1109/itsc.2017.8317941. 8, 61

[68] Saleh, K., Hossny, M., and Nahavandi, S. (2018). Intent prediction of pedestrians via motion trajectories using stacked recurrent neural networks. *IEEE Transactions on Intelligent Vehicles*, 3(4):414–424. DOI: 10.1109/tiv.2018.2873901. 8

[69] Vicente, F., Huang, Z., Xiong, X., et al. (2015). Driver gaze tracking and eyes off the road detection system. *IEEE Transactions on Intelligent Transportation Systems*, 16(4):2014–2027. DOI: 10.1109/tits.2015.2396031. 9

[70] Yan, C., Coenen, F., and Zhang, B. (2014). Driving posture recognition by joint application of motion history image and pyramid histogram of oriented gradients. *International Journal of Vehicular Technology*, 719413:1–11. DOI: 10.1155/2014/719413. 9

[71] Liang, Y., Lee, J. D., and Reyes, M. L. (2007). Nonintrusive detection of driver cognitive distraction in real time using Bayesian networks. *Transportation Research Record*, 2008(1):1–8. DOI: 10.3141/2018-01. 9

[72] Liang, Y., Reyes, M. L., and Lee, J. D. (2007). Real-time detection of driver cognitive distraction using support vector machines. *IEEE Transactions on Intelligent Transportation Systems*, 8(2):340–350. DOI: 10.1109/tits.2007.895298. 9

[73] Tran, C., Doshi, A., and Trivedi, M. M. (2012). Modeling and prediction of driver behavior by foot gesture analysis. *Computer Vision and Image Understanding*, 116(3):435–445. DOI: 10.1016/j.cviu.2011.09.008. 9

[74] Craye, C. and Karray, F. (2015). Driver distraction detection and recognition using RGB-D sensor. *ArXiv Preprint ArXiv:1502.00250.* 9

[75] Braunagel, C., Kasneci, E., Stolzmann, W., et al. (2015). Driver-activity recognition in the context of conditionally autonomous driving. *IEEE 18th International Conference on Intelligent Transportation Systems*, pages 1652–1657. DOI: 10.1109/itsc.2015.268. 9

[76] Almahasneh, H. S., Kamel, N., Malik, A. S., et al. (2014). EEG based driver cognitive distraction assessment. *5th International Conference on Intelligent and Advanced Systems (ICIAS)*, IEEE. DOI: 10.1109/icias.2014.6869460. 9

[77] Rumelhart, D. E., Hinton, G. E., and Williams, R. J. (1986). Learning representations by back propagating errors. *Nature*, 323(6088):533–536. DOI: 10.1038/323533a0. 9

[78] Berg, A., Deng, J., and Fei-Fei, L. (2010). Large scale visual recognition challenge. DOI: 10.1007/s11263-015-0816-y. 10

[79] Krizhevsky, A., Sutskever, I., and Hinton, G. E. (2017). ImageNet classification with deep convolutional neural networks. *Communications of the ACM*, 60(6):84–90. DOI: 10.1145/3065386. 10

[80] LeCun, Y., Boser, B. E., Denker, J. S., et al. (1990). Handwritten digit recognition with a back-propagation network. *Proc. of Advances in Neural Information Processing Systems*, pages 396–404. 10

[81] He, K., Zhang, X., Ren, S., and Sun, J. (2016). Deep residual learning for image recognition. *Proc. of the IEEE Conference on Computer Vision and Pattern Recognition*, pages 770–778. DOI: 10.1109/cvpr.2016.90. 10

[82] Okon, O. D. and Meng, L. (2017). Detecting distracted driving with deep learning. *International Conference on Interactive Collaborative Robotics*, pages 170–179, Springer, Cham. DOI: 10.1007/978-3-319-66471-2_19. 10

[83] Yan, S., Teng, Y., Smith, J. S., et al. (2016). Driver behavior recognition based on deep convolutional neural networks. *12th International Conference on Natural Computation, Fuzzy Systems and Knowledge Discovery (ICNC-FSKD)*, pages 636–641, IEEE. DOI: 10.1109/fskd.2016.7603248. 10

[84] Tran, D., Do, H. M., Sheng, W., et al. (2018). Real-time detection of distracted driving based on deep learning. *IET Intelligent Transport Systems*, 12(10):1210–1219. DOI: 10.1049/iet-its.2018.5172. 10

[85] Eraqi, H. M., Abouelnaga, Y., Saad, M. H., and Moustafa, M. N. (2019). Driver distraction identification with an ensemble of convolutional neural networks. *Journal of Advanced Transportation*, 2019. DOI: 10.1155/2019/4125865. 10

[86] Martin, M., Popp, J., Anneken, M., et al. (2018). Body pose and context information for driver secondary task detection. *IEEE Intelligent Vehicles Symposium (IV)*, pages 2015–2021. DOI: 10.1109/ivs.2018.8500523. 10

[87] Olabiyi, O., Martinson, E., Chintalapudi, V., et al. (2017). Driver action prediction using deep (bidirectional) recurrent neural network. *ArXiv Preprint ArXiv:1706.02257*. 10

[88] Peng, X., Liu, R., Murphey, Y. L., et al. (2018). Driving maneuver detection via sequence learning from vehicle signals and video images. *24th International Conference on Pattern Recognition (ICPR)*, pages 1265–1270, IEEE. DOI: 10.1109/icpr.2018.8546255. 10

[89] Weyers, P., Schiebener, D., and Kummert, A. (2019). Action and object interaction recognition for driver activity classification. *IEEE Intelligent Transportation Systems Conference (ITSC)*. DOI: 10.1109/itsc.2019.8917139. 11

[90] Liu, J., Wang, G., Duan, L. Y., et al. (2017). Skeleton-based human action recognition with global context-aware attention LSTM networks. *IEEE Transactions on Image Processing*, 27(4):1586–1599. DOI: 10.1109/tip.2017.2785279. 11

[91] Zhang, S., Liu, X., and Xiao, J. (2017). On geometric features for skeleton-based action recognition using multilayer LSTM networks. *IEEE Winter Conference on Applications of Computer Vision (WACV)*, pages 148–157. DOI: 10.1109/wacv.2017.24. 11

[92] Yan, S., Xiong, Y., and Lin, D. (2018). Spatial temporal graph convolutional networks for skeleton-based action recognition. *Proc. of the AAAI Conference on Artificial Intelligence*. 11

[93] Elander, J., West, R., and French, D. (1993). Behavioral correlates of individual differences in road-traffic crash risk: An examination of methods and findings. *Psychological Bulletin*, 113(2). DOI: 10.1037/0033-2909.113.2.279. 11

[94] Berry, I. M. (2010). The effects of driving style and vehicle performance on the real-world fuel consumption of U.S. light-duty vehicles. Dissertation, Massachusetts Institute of Technology. 11

[95] Bingham, C., Walsh, C., and Carroll, S. (2012). Impact of driving characteristics on electric vehicle energy consumption and range. *IET Intelligent Transport Systems*, 6(1):29–35. DOI: 10.1049/iet-its.2010.0137. 11

[96] Felipe, J., Amarillo, J. C., Naranjo, J. E., et al. (2015). Energy consumption estimation in electric vehicles considering driving style. *IEEE 18th International Conference on Intelligent Transportation Systems*, pages 101–106. DOI: 10.1109/itsc.2015.25. 11

[97] Wu, X., Freese, D., Cabrera, A., et al. (2015). Electric vehicles' energy consumption measurement and estimation. *Transportation Research Part D: Transport and Environment*, 34:52–67. DOI: 10.1016/j.trd.2014.10.007. 11

[98] Ouali, T., Shah, N., Kim, B., et al. (2016). Driving style identification algorithm with real-world data based on statistical approach. *SAE Technical Paper. (no. 2016–01-1422)*. DOI: 10.4271/2016-01-1422. 11

[99] Constantinescu, Z., Marinoiu, C., and Vladoiu, M. (2010). Driving style analysis using data mining techniques. *International Journal of Computers Communications and Control*, 5(5):654–663. DOI: 10.15837/ijccc.2010.5.2221. 11

[100] Hartwich, F., Beggiato, M., Krems, J. F. (2018). Driving comfort, enjoyment and acceptance of automated driving—effects of drivers' age and driving style familiarity. *Ergonomics*, 61(8):1017–1032. DOI: 10.1080/00140139.2018.1441448. 11

[101] Basu, C., Yang, Q., Hungerman, D., et al. (2017). Do you want your autonomous car to drive like you? *12th ACM/IEEE International Conference on Human-Robot Interaction*, pages 417–425. DOI: 10.1145/2909824.3020250. 11, 12, 91

[102] Kuderer, M., Gulati, S., and Burgard, W. (2015). Learning driving styles for autonomous vehicles from demonstration. *IEEE International Conference on Robotics and Automation (ICRA)*, pages 2641–2646. DOI: 10.1109/icra.2015.7139555. 11, 100

[103] Ishibashi, M., Okuwa, M., Doi, S., et al. (2007). Indices for characterizing driving style and their relevance to car following behavior. *SICE Annual Conference*, pages 1132–1137, IEEE. DOI: 10.1109/sice.2007.4421155. 12

[104] Wiesenthal, D. L., Hennessy, D., and Gibson, P. M. (2000). The driving vengeance questionnaire (DVQ): The development of a scale to measure deviant drivers' attitudes. *Violence and Victims*, 15(2):115–136. DOI: 10.1891/0886-6708.15.2.115. 12

[105] Dula, C. S. and Ballard, M. E. (2003). Development and evaluation of a measure of dangerous, aggressive, negative emotional, and risky driving. *Journal of Applied Social Psychology*, 33(2):263–282. DOI: 10.1111/j.1559-1816.2003.tb01896.x. 12

[106] French, D. J., West, R. J., Elander, J., et al. (1993). Decision-making style, driving style, and self-reported involvement in road traffic accidents. *Ergonomics*, 36(6):627–644. DOI: 10.1080/00140139308967925. 12

[107] Taubman-Ben-Ari, O., Mikulincer, M., and Gillath, O. (2004). The multidimensional driving style inventory—scale construct and validation. *Accident Analysis and Prevention*, 36(3):323–332. DOI: 10.1016/s0001-4575(03)00010-1. 12

[108] Ishibashi, M., Okuwa, M., Doi, S. I., and Akamatsu, M. (2007). Indices for characterizing driving style and their relevance to car following behavior. *IEEE*, pages 1132–1137. DOI: 10.1109/SICE.2007.4421155. 12

[109] Vaitkus, V., Lengvenis, P., and Žylius, G. (2014). Driving style classification using long-term accelerometer information. *19th International Conference on Methods and Models in Automation and Robotics (MMAR)*, pages 641–644, IEEE. DOI: 10.1109/mmar.2014.6957429. 12, 94

[110] Murphey, Y. L., Milton, R., and Kiliaris, L. (2009). Driver's style classification using jerk analysis. *IEEE Workshop on Computational Intelligence in Vehicles and Vehicular Systems*, pages 23–28. DOI: 10.1109/civvs.2009.4938719. 12, 95

[111] Miyajima, C., Nishiwaki, Y., Ozawa, K., Wakita, T., Itou, K., Takeda, K., and Itakura, F. (2007). Driver modeling based on driving behavior and its evaluation in driver identification. *Proc. of the IEEE*, 95(2):427–437. DOI: 10.1109/jproc.2006.888405. 12

[112] Shi, B., Xu, L., Hu, J., et al. (2015). Evaluating driving styles by normalizing driving behavior based on personalized driver modeling. *IEEE Transactions on Systems, Man, and Cybernetics: Systems*, 45(12):1502–1508. DOI: 10.1109/tsmc.2015.2417837. 12

[113] Chen, K. T. and Chen, H. Y. W. (2019). Driving style clustering using naturalistic driving data. *Transportation Research Record*, 2673(6):176–188. DOI: 10.1177/0361198119845360. 12, 100

[114] Olsen, E. C., Lee, S. E., and Simons-Morton, B. G. (2007). Eye movement patterns for novice teen drivers: Does 6 months of driving experience make a difference? *Transportation Research Record*, 2009(1):8–14. DOI: 10.3141/2009-02. 12

[115] Sarkar, S. and Andreas, M. (2004). Acceptance of and engagement in risky driving behaviors by teenagers. *Adolescence*, 39(156):687. 12

[116] Constantinou, E., Panayiotou, G., Konstantinou, N., Loutsiou-Ladd, A., and Kapardis, A. (2011). Risky and aggressive driving in young adults: Personality matters. *Accident Analysis and Prevention*, 43(4):1323–1331. DOI: 10.1016/j.aap.2011.02.002. 12, 13, 14, 112

[117] Rhodes, N. and Pivik, K. (2011). Age and gender differences in risky driving: The roles of positive affect and risk perception. *Accident Analysis and Prevention*, 43(3):923–931. DOI: 10.1016/j.aap.2010.11.015. 128

[118] Evans, L. and Wasielewski, P. (1982). Do accident-involved drivers exhibit riskier everyday driving behavior? *Accident Analysis and Prevention*, 14(1):57–64. DOI: 10.1016/0001-4575(82)90007-0. 13

[119] Cestac, J., Paran, F., and Delhomme, P. (2011). Young drivers' sensation seeking, subjective norms, and perceived behavioral control and their roles in predicting speeding intention: How risk-taking motivations evolve with gender and driving experience. *Safety Science*, 49(3):424–432. DOI: 10.1016/j.ssci.2010.10.007. 13

[120] Harbeck, E. L. and Glendon, A. I. (2018). Driver prototypes and behavioral willingness: Young driver risk perception and reported engagement in risky driving. *Journal of Safety Research*, 66:195–204. DOI: 10.1016/j.jsr.2018.07.009. 13, 113

[121] Zuckerman, M. (1971). Dimensions of sensation seeking. *Journal of Consulting and Clinical Psychology*, 36(1):45. DOI: 10.1037/h0030478. 13

[122] Burns, P. C. and Wilde, G. J. (1995). Risk taking in male taxi drivers: Relationships among personality, observational data, and driver records. *Personality and Individual Differences*, 18(2):267–278. DOI: 10.1016/0191-8869(94)00150-q. 13

[123] Zuckerman, M. (1994). Behavioral expressions and biosocial bases of sensation seeking. Cambridge University Press. 13

[124] Rundmo, T. and Iversen, H. (2004). Risk perception and driving behaviour among adolescents in two Norwegian counties before and after a traffic safety campaign. *Safety Science*, 42(1):1–21. DOI: 10.1016/s0925-7535(02)00047-4. 13

[125] Deery, H. A. (1999). Hazard and risk perception among young novice drivers. *Journal of Safety Research*, 30(4):225–236. DOI: 10.1016/S0022-4375(99)00018-3. 14

[126] Jonah, B. A., Thiessen, R., and Au-Yeung, E. (2001). Sensation seeking, risky driving, and behavioral adaptation. *Accident Analysis and Prevention*, 33(5):679–684. DOI: 10.1016/s0001-4575(00)00085-3. 14

[127] Ulleberg, P. and Rundmo, T. (2003). Personality, attitudes, and risk perception as predictors of risky driving behaviour among young drivers. *Safety Science*, 41(5):427–443. DOI: 10.1016/s0925-7535(01)00077-7. 14, 130, 139

[128] Zhang, L., Zhang, C., and Shang, L. (2016). Sensation-seeking and domain-specific risk-taking behavior among adolescents: Risk perceptions and expected benefits as mediators. *Personality and Individual Differences*, 101:299–305. DOI: 10.1016/j.paid.2016.06.002. 14, 113

[129] Eason, J., Manstead, A., Stradling, S., Baxter, J., and Campbell, K. (1990). Errors and violations on the roads: A real distinction? *Ergonomics*, 33(10–11):1315–1332. DOI: 10.1080/00140139008925335. 14

[130] Bingham, C. R. and Ehsani, J. P. (2012). The relative odds of involvement in seven crash configurations by driver age and sex. *Journal of Adolescent Health*, 51(5):484–490. DOI: 10.1016/j.jadohealth.2012.02.012. 14

[131] Norris, F. H., Matthews, B. A., and Riad, J. K. (2000). Characterological, situational, and behavioral risk factors for motor vehicle accidents: A prospective examination. *Accident Analysis and Prevention*, 32:505–515. DOI: 10.1016/s0001-4575(99)00068-8. 14

[132] U.S. Department of Transportation. Next-Generation Simulation (NGSIM) Vehicle Trajectories and Supporting Data. https://data.transportation.gov/Automobiles/Next-Generation-Simulation-NGSIM-Vehicle-Trajector/8ect-6jqj, 2020-06-23. 17, 91

[133] Alexiadis, V., Colyar, J., Halkias, J., et al. (2004). The next-generation simulation program. *Institute of Transportation Engineers. ITE Journal*, 74(8):22. 17

[134] Press, W. H. and Teukolsky, S. A. (1990). Savitzky-Golay smoothing filters. *Computers in Physics*, 4(6):669–672. DOI: 10.1063/1.4822961. 17

[135] Fang, H. S., Xie, S., Tai, Y. W., et al. (2017). RMPE: Regional multi-person pose estimation. *IEEE International Conference on Computer Vision (ICCV)*. DOI: 10.1109/iccv.2017.256. 41

[136] Zhe, C., Simon, T., Wei, S. E., et al. (2017). Realtime multi-person 2D pose estimation using part affinity fields. *IEEE Conference on Computer Vision and Pattern Recognition (CVPR)*. DOI: 10.1109/cvpr.2017.143. 41

[137] Yano, S., Gu, Y., and Kamijo, S. (2016). Estimation of pedestrian pose and orientation using on-board camera with histograms of oriented gradients features. *International Journal of Intelligent Transportation Systems Research*. DOI: 10.1007/s13177-014-0103-2. 43

[138] Schulz, A. and Stiefelhagen, R. (2012). Video-based pedestrian head pose estimation for risk assessment. *15th International IEEE Conference on Intelligent Transportation Systems*, pages 1771–1776. DOI: 10.1109/itsc.2012.6338829. 43

[139] Grewal, M. S. and Andrews, A. P. (2014). *Kalman Filtering: Theory and Practice with MATLAB*. John Wiley & Sons. DOI: 10.1002/9780470377819. 52

[140] Mazor, E., Averbuch, A., Bar-Shalom, Y., et al. (1998). Interacting multiple model methods in target tracking: A survey. *IEEE Transactions on Aerospace and Electronic Systems*, 34(1):103–123. DOI: 10.1109/7.640267. 54

[141] Peng, Z., Wei, S., Tian, J., et al. (2016). Attention-based bidirectional long short-term memory networks for relation classification. *Proc. of the 54th Annual Meeting of the Association for Computational Linguistics, Volume 2: Short Papers*. DOI: 10.18653/v1/P16-2034. 59

[142] Yang, Y., Yang, T., Wang, K. L., et al. (2014). Vehicle active collision avoidance system. *Acta Sichuan Armory Engineering*, 35(10):77–81. DOI: 10.4271/2019-01-0130. 61

[143] Zhu, M., Wang, X., Tarko, A., and Fang, S. (2018). Modeling car-following behavior on urban expressways in Shanghai: A naturalistic driving study. *Transportation Research. Part C: Emerging Technology*, 93:425–445. DOI: 10.1016/j.trc.2018.06.009. 92

[144] Jacoby, W. G. (2000). Loess: A nonparametric, graphical tool for depicting relationships between variables. *Electoral Studies*, 19(4):577–613. DOI: 10.1016/S0261-3794(99)00028-1. 93

[145] Mahmud, S. S., Ferreira, L., Hoque, M. S., and Tavassoli, A. (2017). Application of proximal surrogate indicators for safety evaluation: A review of recent developments and research needs. *IATSS Research*, 41(4):153–163. DOI: 10.1016/j.iatssr.2017.02.001. 94

[146] Gan, G. and Ng, M. K. P. (2017). *K*-means clustering with outlier removal. *Pattern Recognition Letters*, 90:8–14. DOI: 10.1016/j.patrec.2017.03.008. 97

[147] Black, K. (2019). *Business Statistics: For Contemporary Decision Making*, John Wiley & Sons. 99

[148] Bentler, P. M. and Chou, C. P. (1987). Practical issues in structural modeling. *Sociological Methods and Research*, 16(1):78–117. DOI: 10.1177/0049124187016001004. 112

[149] Kline, R. B. (2015). *Principles and Practice of Structural Equation Modeling*, Guilford Publications. 112, 113

[150] Cordazzo, S. T., Scialfa, C. T., Bubric, K., and Ross, R. J. (2014). The driver behaviour questionnaire: A North American analysis. *Journal of Safety Research*, 50:99–107. DOI: 10.1016/j.jsr.2014.05.002. 113

[151] Parker, D., Reason, J. T., Manstead, A. S., and Stradling, S. G. (1995). Driving errors, driving violations, and accident involvement. *Ergonomics*, 38(5):1036–1048. DOI: 10.1080/00140139508925170. 113

[152] Lawton, R., Parker, D., Manstead, A. S., and Stradling, S. G. (1997). The role of affect in predicting social behaviors: The case of road traffic violations. *Journal of Applied Social Psychology*, 27(14):1258–1276. DOI: 10.1111/j.1559-1816.1997.tb01805.x. 113

[153] Zhao, N., Mehler, B., Reimer, B., D'Ambrosio, L. A., Mehler, A., and Coughlin, J. F. (2012). An investigation of the relationship between the driving behavior questionnaire and objective measures of highway driving behavior. *Transportation Research Part F: Traffic Psychology and Behaviour*, 15(6):676–685. DOI: 10.1016/j.trf.2012.08.001. 113

[154] Blockey, P. N. and Hartley, L. R. (1995). Aberrant driving behaviour: Errors and violations. *Ergonomics*, 38:1759–1771. DOI: 10.1080/00140139508925225. 113

[155] Lajunen, T., Parker, D., and Summala, H. (2004). The Manchester driver behaviour questionnaire: A cross-cultural study. *Accident Analysis and Prevention*, 36(2):231–238. DOI: 10.1016/s0001-4575(02)00152-5. 113

[156] Hamdar, S. H., Mahmassani, H. S., and Chen, R. B. (2008). Aggressiveness propensity index for driving behavior at signalized intersections. *Accident Analysis and Prevention*, 40(1):315–326. DOI: 10.1016/j.aap.2007.06.013. 113, 123

[157] Lee, J. Y., Chung, J. H., and Son, B. (2008). Analysis of traffic accident size for Korean highway using structural equation models. *Accident Analysis and Prevention*, 40(6):1955–1963. DOI: 10.1016/j.aap.2008.08.006. 123

[158] Newsom, J. T. (2012). Some clarifications and recommendations on fit indices. *USP*, 655:123–133. 124

[159] Byrne, B. M. (2001). Structural equation modeling: Perspectives on the present and the future. *International Journal of Testing*, 1(3–4):327–334. 126

[160] De Winter, J. C. F. and Dodou, D. (2010). The driver behaviour questionnaire as a predictor of accidents: A meta-analysis. *Journal of Safety Research*, 41(6):463–470. DOI: 10.1016/j.jsr.2010.10.007. 128

[161] Nordfjærn, T., Jørgensen, S. H., and Rundmo, T. (2010). An investigation of driver attitudes and behaviour in rural and urban areas in Norway. *Safety Science*, 48(3):348–356. DOI: 10.1016/j.ssci.2009.12.001. 129

[162] Delhomme, P., Verlhiac, J., and Martha, C. (2009). Are drivers' comparative risk judgments about speeding realistic? *Journal of Safety Research*, 40(5):333–339. DOI: 10.1016/j.jsr.2009.09.003. 129

[163] Machin, M. A. and Sankey, K. S. (2008). Relationships between young drivers' personality characteristics, risk perceptions, and driving behaviour. *Accident Analysis and Prevention*, 40(2):541–547. DOI: 10.1016/j.aap.2007.08.010. 129

[164] Mills, B., Reyna, V. F., and Estrada, S. (2008). Explaining contradictory relations between risk perception and risk taking. *Psychological Science*, 19(5):429–433. DOI: 10.1111/j.1467-9280.2008.02104.x. 129

[165] Smiley, A. (2004). Adaptive strategies of older drivers. *Transportation Research Board Conference Proceedings*, 27. 132

[166] Antin, J. F., Guo, F., Fang, Y., Dingus, T. A., Hankey, J. M., and Perez, M. A. (2017). The influence of functional health on seniors' driving risk. *Journal of Transport and Health*, 6:237–244. DOI: 10.1016/j.jth.2017.07.003. 132

[167] McGwin, Jr. G. and Brown, D. B. (1999). Characteristics of traffic crashes among young, middle-aged, and older drivers. *Accident Analysis and Prevention*, 31(3):181–198. DOI: 10.1016/s0001-4575(98)00061-x. 132

[168] Bauer, M. J., Adler, G., Kuskowski, M. A., and Rottunda, S. (2003). The influence of age and gender on the driving patterns of older adults. *Journal of Women and Aging*, 15(4):3–16. DOI: 10.1300/j074v15n04_02. 132

[169] Molnar, L. J., Eby, D. W., Charlton, J. L., Langford, J., Koppel, S., Marshall, S., and Man-Son-Hing, M. (2013). Reprint of driving avoidance by older adults: Is it always self-regulation? *Accident Analysis and Prevention*, 61:272–280. DOI: 10.1016/j.aap.2013.07.004. 132

[170] D'Ambrosio, L. A., Donorfio, L. K., Coughlin, J. F., Mohyde, M., and Meyer, J. (2008). Gender differences in self-regulation patterns and attitudes toward driving among older adults. *Journal of Women and Aging*, 20(3–4):265–282. DOI: 10.1080/08952840801984758. 132

[171] Guo, F. and Fang, Y. (2013). Individual driver risk assessment using naturalistic driving data. *Accident Analysis and Prevention*, 61:3–9. DOI: 10.1016/j.aap.2012.06.014. 133, 138

[172] Safavian, S. R. and Landgrebe, D. (1991). A survey of decision tree classifier methodology. *IEEE Transactions on Systems, Man, and Cybernetics*, 21(3):660–674. DOI: 10.1109/21.97458. 134

[173] Breiman, L. (1996). Bagging predictors. *Machine Learning*, 24(2):123–140. DOI: 10.1007/bf00058655. 135

[174] Bagdadi, O. (2013). Assessing safety critical braking events in naturalistic driving studies. *Transportation Research Part F: Traffic Psychology and Behaviour*, 16:117–126. DOI: 10.1016/j.trf.2012.08.006. 135

[175] Zheng, L., Ismail, K., and Meng, X. (2014). Freeway safety estimation using extreme value theory approaches: A comparative study. *Accident Analysis and Prevention*, 62:32–41. DOI: 10.1016/j.aap.2013.09.006. 135

[176] Dingus, T. A., Klauer, S. G., Neale, V. L., Petersen, A., Lee, S. E., Sudweeks, J., and Bucher, C. (2006). The 100-car naturalistic driving study, phase II-results of the 100-car field experiment (no. DOT-HS-810–593). The United States. Department of Transportation. National Highway Traffic Safety Administration. DOI: 10.1037/e624282011-001. 135

[177] Wang, X. and Xu, X. (2019). Assessing the relationship between self-reported driving behaviors and driver risk using a naturalistic driving study. *Accident Analysis and Prevention*, 128:8–16. DOI: 10.1016/j.aap.2019.03.009. 135

[178] Xing, Y., Lv, C., Zhang, Z., Cao, D., Velenis, E., and Wang, F. (2019). Driver activity recognition for intelligent vehicles: A deep learning approach. *IEEE Transactions on Vehicular Technology*, 68(6):5379–5390. DOI: 10.1109/TVT.2019.2908425.

[179] Lyons, T. J., Kenworthy, J. R., and Newman, P. W. G. (1990). Urban structure and air pollution. *Atmospheric Environment. Part B. Urban Atmosphere*, 24(1):43–48. DOI: 10.1016/0957-1272(90)90008-I. 101

[180] Sysoev, M., Kos, A., Guna, J., et al. (2017). Estimation of the driving style based on the users' activity and environment influence. *Sensors*, 17(10):2404. DOI: 10.3390/s17102404. 101

[181] Faria, M., Duarte, G., Varella, R., Farias, T., and Baptista, P. (2019). How to road grade, road type and driving aggressiveness impact vehicle fuel consumption? Assessing potential fuel savings in Lisbon, Portugal. *Transportation Research Part D—Transport and Environment*, 72:148–161. DOI: 10.1016/j.trd.2019.04.016. 102

Authors' Biographies

XIAOLIN SONG

Xiaolin Song received her B.E., M.E., and Ph.D. at the College of Mechanical and Vehicle Engineering, Hunan University in 1988, 1991, and 2007, respectively. From 2008 to the present, she has been a professor and a Ph.D. supervisor at Hunan University. She was an advanced visiting scholar of the University of Michigan (Ann Arbor), the University of Waterloo, and the University of Texas at Austin. She is a Vice-Chairman of the Rules Committee of Formula Student China, as well as an Academic Committee Member of the College of Mechanical and Vehicle Engineering, Hunan University. She has been an independent Principal Investigator (PI) and Co-PIs for multiple General Projects of the Natural Science Foundation of China (NFSC) and Hunan Provincial Natural Science Foundation, and dozens of other provincial and ministerial projects or industrial companies. Her research interests include the active safety of intelligent vehicles, vehicle dynamics control, driver behavior modeling, and human factors in driving safety.

HAOTIAN CAO

Haotian Cao is currently a Postdoctoral Fellow at the College of Mechanical and Vehicle Engineering, Hunan University, Changsha, China. He received a B.E. in vehicle engineering and a Ph.D. in mechanical engineering from the College of Mechanical and Vehicle Engineering, Hunan University, Changsha, China, in 2011 and 2018, respectively. He was a visiting scholar at the Human Factors group, the University of Michigan Transportation Research Institute (UMTRI) from 2016–2017. He is currently a committee member of the Chinese Association of Automation Parallel Intelligence (2018–2022), and a referee of over 30 international journals and conferences. He is also the Principal Investigator of Youth Projects funded by the Natural Science Foundation of China (NFSC) and Hunan Provincial Natural Science Foundation, as well as Co-PIs for several projects from the NFSC and industry companies. His interests include trajectory planning and following control for intelligent vehicles, driver models, driver behavior modeling, and naturalistic driving data analysis.

Printed in the United States
by Baker & Taylor Publisher Services